# THIS BOOK BELONGS TO

.................................................

Copyright © 2023 by HarperCollinsPublishers Limited

All rights reserved. Published in the United States by Bright Matter Books, an imprint of Random House Children's Books, a division of Penguin Random House LLC, New York. Originally published in the United Kingdom by Red Shed, part of Farshore, an imprint of HarperCollinsPublishers, London, in 2023.

Bright Matter Books and colophon are registered trademarks of Penguin Random House LLC.

Visit us on the Web! rhcbooks.com

Educators and librarians, for a variety of teaching tools, visit us at RHTeachersLibrarians.com

Library of Congress Cataloging-in-Publication Data is available upon request.
ISBN 978-0-593-90335-3 (trade) — ISBN 978-0-593-90336-0 (ebook)

Written by Miranda Smith
Interior illustrations by Kaja Kajfež, Santiago Calle, Mateo Markov, and Max Rambaldi
Front cover illustrations by Kaja Kajfež, Mateo Markov, and Santiago Calle
Consultancy by Dr. Ashwini V. Mohan
Interior design by Duck Egg Blue Limited

MANUFACTURED IN MALAYSIA
10 9 8 7 6 5 4 3 2 1

First U.S. Edition

Random House Children's Books supports the First Amendment and celebrates the right to read.

Penguin Random House LLC supports copyright. Copyright fuels creativity, encourages diverse voices, promotes free speech, and creates a vibrant culture. Thank you for buying an authorized edition of this book and for complying with copyright laws by not reproducing, scanning, or distributing any part in any form without permission. You are supporting writers and allowing Penguin Random House to publish books for every reader.

# AN ANIMAL A DAY

WRITTEN BY
MIRANDA SMITH

ILLUSTRATED BY
KAJA KAJFEŽ,
SANTIAGO CALLE,
MATEO MARKOV
AND MAX RAMBALDI

BRIGHT MATTER BOOKS

# CONTENTS

A world of animals   **10**
All about animals   **12**

### January   14
Hunting methods   20
Record breakers   28

### February   30
Living with a volcano   36
Animals about the house   42

### March   46
In the rainforest   50
Claws and talons   56

### April   62
Deep sea mysteries   68
Mimicry   76

### May   78
Migration   84
Desert survival   90

### June   94
Life in freshwater   98
Home building   104

### July   110
Jaws and teeth   116
Flight   122

### August   126
Tree living   132
On the coral reef   140

### September   142
Animal assassins   146
Surviving the cold   154

### October   158
Nightlife   164
Dormancy   174

### November   176
Urban animals   180
Out of sight   186

### December   192
Animal superpowers   196

In danger   210
Conservation success   212
Quiz   214
Glossary   216
Index   218

# A WORLD OF ANIMALS

Animals flourish everywhere on Earth – from the deepest ocean to the highest mountain, from deserts and jungles to meadows and mangrove swamps, from kelp forests to grasslands. This is your animal calendar, with an awesome animal for every day of the year. Discover an animal a day for yourself – which one is on your birthday? Or share that day's creature with your family, friends, carers and teachers, before exploring the rest of the book to find out more about the animals that share our planet.

# ALL ABOUT ANIMALS

An animal is a living thing that gets its energy from food and senses what is happening around it, reacting rapidly to any stimulus. The majority of animals are cold-blooded, which means they cannot produce their own body heat and rely on their surroundings to regulate their temperature. Warm-blooded animals produce their own body heat even when it is cold outside. Animals that are alike are grouped together in classifications of vertebrates and invertebrates.

## VERTEBRATES

These are animals that have a backbone inside their body. The five main classifications are: mammals, birds, amphibians, reptiles and fish. The largest animals in the main ecosystems are vertebrates, for example the elephant and the blue whale.

## MAMMALS

You are a mammal. So are many of the animals in the world. Mammals are warm-blooded, have fur or hair, and feed their young with milk.

American bison
(see p.199)

Townsend's big-eared bat
(see p.148)

## BIRDS

These are the only animals that have feathers. Most of them can fly but some are flightless. Birds are warm-blooded and lay hard-shelled eggs.

Bald eagle
(see p.103)

Ostrich
(see p.166)

## REPTILES

Covered in protective scales or bony plates, these animals need to live in warm habitats. Reptiles are cold-blooded and usually lay soft-shelled eggs.

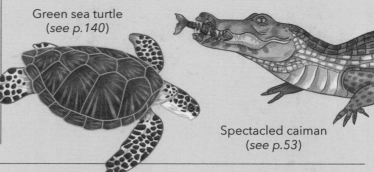

Green sea turtle
(see p.140)

Spectacled caiman
(see p.53)

## AMPHIBIANS

Amphibians need water or damp conditions to survive, and live on land as well as in the water. They are cold-blooded and lay jelly-covered eggs.

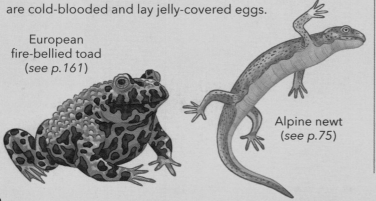

European fire-bellied toad
(see p.161)

Alpine newt
(see p.75)

## FISH

Most fish have scales and use their fins to swim in saltwater or freshwater, and sometimes both. Fish are cold-blooded and breathe through gills.

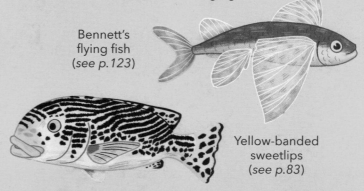

Bennett's flying fish
(see p.123)

Yellow-banded sweetlips
(see p.83)

# INVERTEBRATES

At least 97 percent of all animal species on Earth are invertebrates. They do not have a backbone. They either have a soft body, like a worm or jellyfish, or a hard outer casing called an exoskeleton that covers their body, for example a crab or beetle.

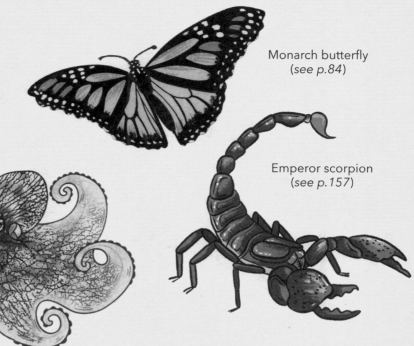

Monarch butterfly (see p.84)

Emperor scorpion (see p.157)

Short-horned grasshopper (see p.108)

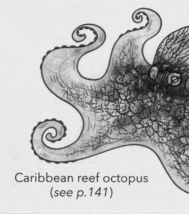

Caribbean reef octopus (see p.141)

# A WORLD OF DISCOVERY

New species of animals are being discovered all the time. For example, in May 2023, scientists announced that more than 5,000 new species had been found living on the seabed of the Clarion-Clipperton Zone, an unexplored area of the Pacific Ocean. And in the same month, 90 new species from the Mekong region in southeastern Asia were announced.

The Cambodian blue-crested agama (*Calotes goetzi*), a newly discovered species

# TRACKING ANIMAL POPULATIONS

The International Union for Conservation of Nature (IUCN) is the international partnership of countries, agencies and other organizations that assesses the conservation status of species of animals. In the data section by every entry in this book, we have included the IUCN's Red List conservation rating. The Red List checks which animals are in danger of extinction and rates them as follows:

**Critically Endangered** – at extremely high risk of extinction in the wild
**Endangered** – at very high risk of extinction in the wild
**Vulnerable** – at high risk of extinction in the wild
**Near Threatened** – close to qualifying for a threatened category in the near future
**Least Concern** – population is stable
**Data Deficient** – there is not enough information to assess risk
**Not Evaluated** – the species has not yet been studied

Beluga sturgeon are Critically Endangered (see p.99).

# JANUARY

• January 1st •
## POLAR BEAR

A mother and her two cubs make their way across the melting sea ice that covers the waters around the Arctic Circle. She gave birth in her winter snow cave and now the cubs are big enough to be shown how to swim, hunt seals and survive in the cold. They will stay with their devoted mother until they are around three years old.

| | |
|---|---|
| **SCIENTIFIC NAME** | *Ursus maritimus* |
| **ANIMAL GROUP** | Mammals |
| **LENGTH/WEIGHT** | up to 10 ft. including tail/1,500 lb. |
| **DIET** | carnivore: seals, Arctic foxes |
| **LOCATION** | Arctic |
| **STATUS** | Vulnerable |

### • January 2nd •
## TUNKI

This colorful male bird, also known as the Andean cock-of-the-rock, is the national bird of Peru. It lives in the cloud forests of the Andes Mountains where it shows off its plumage to attract a mate. It raises its crest and confronts competing males with squawks and grunts. A nest is built by the female from mud mixed with saliva and plastered to a rock or cliff. She will raise the young by herself.

| | |
|---|---|
| **SCIENTIFIC NAME** | *Rupicola peruvianus* |
| **ANIMAL GROUP** | Birds |
| **WINGSPAN/WEIGHT** | up to 20 in./11 oz. |
| **DIET** | omnivore: fruit, insects, amphibians, reptiles, mice |
| **LOCATION** | western South America |
| **STATUS** | Least Concern |

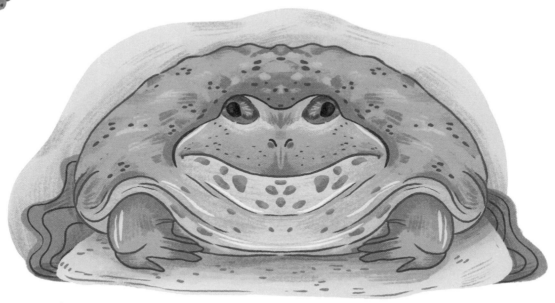

### • January 3rd •
## GOLIATH FROG

Meet the largest frog in the world, weighing about the same as a pet cat. Despite the fact that it looks as if it cannot move, it can jump an amazing 10 feet in one leap, swim against strong currents and lift heavy stones out of the way when building a nest. It does not have a vocal sac, so cannot croak like other frogs.

| | |
|---|---|
| **SCIENTIFIC NAME** | *Conraua goliath* |
| **ANIMAL GROUP** | Amphibians |
| **LENGTH/WEIGHT** | up to 13 in./7 lb. |
| **DIET** | omnivore: insects, crustaceans, fish |
| **LOCATION** | western Africa |
| **STATUS** | Endangered |

• January 4th •
# BIGFIN REEF SQUID

If threatened, this mollusk can cover a distance of 50 feet in only one second! It does this by jet propulsion – taking in water and pushing it out of its body under pressure through a siphon, or tube. It can also protect itself by releasing a dark cloud of ink. If that was not enough, it is a master of disguise and can change its skin cells to a variety of colors to camouflage itself.

| | |
|---|---|
| **SCIENTIFIC NAME** | *Sepioteuthis lessoniana* |
| **ANIMAL GROUP** | Invertebrates |
| **LENGTH/WEIGHT** | up to 13 in./3.1 lb. |
| **DIET** | carnivore: fish, crustaceans |
| **LOCATION** | Indian and western Pacific Oceans |
| **STATUS** | Data Deficient |

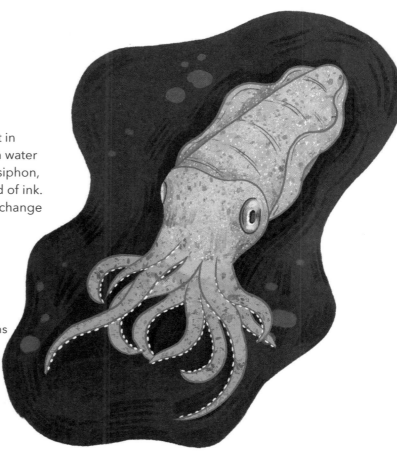

• January 5th •
# FISHING CAT

Well-adapted to swimming and diving, this cat has partially webbed front feet and long guard hairs that cover a short-hair layer of fur next to the skin to keep water out. It has sharp claw tips that do not retract like those of other cats and can quickly seize the fish it hunts in the marshy thickets and mangrove swamps that it inhabits.

| | |
|---|---|
| **SCIENTIFIC NAME** | *Prionailurus viverrinus* |
| **ANIMAL GROUP** | Mammals |
| **LENGTH/WEIGHT** | up to 3.6 ft. including tail/13 lb. |
| **DIET** | carnivore: fish, frogs |
| **LOCATION** | south and southeastern Asia |
| **STATUS** | Vulnerable |

## • January 6th •
## ARMORED GROUND CRICKET

This cricket is very used to defending itself. Its body is covered in spines and it has a nasty bite. It can produce loud noises by stridulation – rubbing the edges of its wings together. It can even squirt blood out of gaps in its body, and if all else fails, vomit on itself and the predator.

| | |
|---|---|
| **SCIENTIFIC NAME** | *Acanthoplus discoidalis* |
| **ANIMAL GROUP** | Invertebrates |
| **LENGTH** | up to 2 in. |
| **DIET** | omnivore: birds, plants, insects, fruit |
| **LOCATION** | southern Africa |
| **STATUS** | Least Concern |

## • January 7th •
## EURASIAN GREEN WOODPECKER

Unlike most woodpeckers, this bird mostly feeds on the ground, where it forages for invertebrates – particularly ants, which it laps up with its 4-inch-long tongue. It is easily recognized in the air because of its undulating flight and laughing call. It drills into trees to make nest holes and both parents care for the young.

| | |
|---|---|
| **SCIENTIFIC NAME** | *Picus viridis* |
| **ANIMAL GROUP** | Birds |
| **WINGSPAN/WEIGHT** | up to 13 in./8 oz. |
| **DIET** | omnivore: insects (mainly ants), worms, pine seeds, fruit |
| **LOCATION** | Europe |
| **STATUS** | Least Concern |

## • January 8th •
## ZEBRA SHARK

When it is born, this shark has stripes with some spots. The adults have no stripes and are covered with spots. Found around coral reefs in tropical waters, it often rests on the sand during the day but is more active at night, wriggling into narrow crevices after prey.

| | |
|---|---|
| **SCIENTIFIC NAME** | *Stegostoma fasciatum* |
| **ANIMAL GROUP** | Fish |
| **LENGTH/WEIGHT** | up to 11 ft./45 lb. |
| **DIET** | omnivore: fish, crabs, snails |
| **LOCATION** | western Pacific Ocean, Indian Ocean |
| **STATUS** | Endangered |

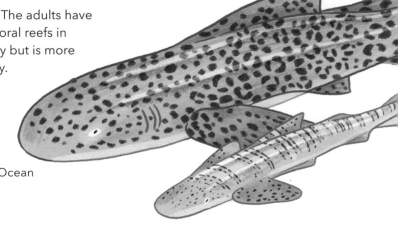

• January 9th •
# PORTUGUESE MAN O' WAR

This is not a single animal. It is a colony of animals called zooids that all work together – some are responsible for floating, some catch prey, some feed, some defend the colony. It drifts on the currents or is blown along by the wind, held on the surface by the zooid at the top – the pneumatophore, a gas-filled bag.

| | |
|---|---|
| **SCIENTIFIC NAME** | *Physalia physalis* |
| **ANIMAL GROUP** | Invertebrates |
| **LENGTH** | tentacles up to 100 ft. long |
| **DIET** | carnivore: small fish and crustaceans |
| **LOCATION** | Indian, Pacific and Atlantic Oceans |
| **STATUS** | Not Evaluated |

# HUNTING METHODS

Predators hunt to survive, providing themselves with the energy their bodies need. To be successful, they must be skillful and agile. Different predators have developed different strategies, some of them very inventive or even unique.

• January 10th •

## OGRE-FACED SPIDER

Hanging from its web, this efficient hunter holds a "capture web" in its front three pairs of legs, ready to throw over any walking or flying prey that passes by. Also known as the gladiator spider, it hunts at night and has the largest eyes of any spider.

| | |
|---|---|
| SCIENTIFIC NAME | Deinopis spinosa |
| ANIMAL GROUP | Invertebrates |
| LENGTH | up to 0.7 in. |
| DIET | carnivore: crickets, beetles, other spiders |
| LOCATION | North America, South America |
| STATUS | Not Evaluated |

• January 11th •

## BANDED ARCHERFISH

Archerfish uniquely shoot down their insect prey. They press their tongues against the roofs of their mouths, close their gill covers, and shoot jets of water at prey up to 7 feet away. They also leap out of the water to snatch insects from the air.

| | |
|---|---|
| SCIENTIFIC NAME | Toxotes jaculatrix |
| ANIMAL GROUP | Fish |
| LENGTH/WEIGHT | up to 12 in./5 lb. |
| DIET | omnivore: insects, fish, plants |
| LOCATION | southeastern Asia, northern Australia |
| STATUS | Least Concern |

• January 12th •

## ELECTRIC EEL

To stun prey, this fish can release an electric charge of up to 800 volts - the supply to houses around the world is usually only up to 240 volts. This eel lives in the muddy waters of the Amazon and Orinoco rivers.

| | |
|---|---|
| SCIENTIFIC NAME | Electrophorus electricus |
| ANIMAL GROUP | Fish |
| LENGTH/WEIGHT | up to 8 ft./45 lb. |
| DIET | omnivore: fish, crustaceans, insects, small amphibians, reptiles, mammals |
| LOCATION | northern South America |
| STATUS | Least Concern |

• January 13th •
# MARGAY

This cat imitates the sound of a young pied tamarin monkey when it is hunting. Scientists say this is the first time a wild cat species has been recorded copying the calls of its prey. The clever hunter can run headfirst down a tree to capture anything it attracts.

| | |
|---|---|
| SCIENTIFIC NAME | *Leopardus wiedii* |
| ANIMAL GROUP | Mammals |
| LENGTH/WEIGHT | up to 4.3 ft. including tail/9 lb. |
| DIET | omnivore: small mammals, birds, reptiles, amphibians, fruit |
| LOCATION | Central and South America |
| STATUS | Near Threatened |

• January 14th •
# ARCTIC FOX

In winter on the tundra, it is difficult to find anything to eat. Arctic foxes use their acute hearing to track small rodents called lemmings underground. Then they leap high in the air and dive into the snow!

| | |
|---|---|
| SCIENTIFIC NAME | *Vulpes lagopus* |
| ANIMAL GROUP | Mammals |
| LENGTH/WEIGHT | up to 3.6 ft. including tail/8 lb. |
| DIET | carnivore: small animals, sea birds, fish, young seals |
| LOCATION | Arctic |
| STATUS | Least Concern |

• January 15th •
# ORCHID MANTIS

Blending perfectly with the rainforest orchid on which it is sitting, the mantis stays immobile, waiting patiently. It is the same color as the orchid and even has petal-shaped legs. When an insect approaches to pollinate the flower, the mantis strikes with lightning precision.

| | |
|---|---|
| SCIENTIFIC NAME | *Hymenopus coronatus* |
| ANIMAL GROUP | Invertebrates |
| LENGTH | up to 2.8 in. |
| DIET | carnivore: crickets, butterflies, moths |
| LOCATION | southeastern Asia |
| STATUS | Vulnerable |

• January 16th •

# LION

When lions hunt big prey, they sometimes work together because it improves their chances of feeding the pride. Some of the lionesses circle round a herd of wildebeest or zebra (see p.195). Then they break cover and chase the frightened animals toward other lionesses, and sometimes male lions, that are hidden in the long grass nearby.

| | |
|---|---|
| **SCIENTIFIC NAME** | *Panthera leo* |
| **ANIMAL GROUP** | Mammals |
| **LENGTH/WEIGHT** | up to 13 ft. including tail/550 lb. |
| **DIET** | carnivore: wildebeest, zebra, gazelles, antelopes, wild hogs |
| **LOCATION** | Africa, south of the Sahara |
| **STATUS** | Vulnerable |

• January 17th •
# BLUE WILDEBEEST

Wildebeest have a tough time on the plains in Africa. They are a favorite prey of lions, cheetahs, hunting dogs and hyenas (*see p.117*). But they are not without defense. They move in large herds, they can run up to 50 mph, and their sharp, curving horns can prove useful when fending off a determined predator.

| | |
|---|---|
| **SCIENTIFIC NAME** | *Connochaetes taurinus* |
| **ANIMAL GROUP** | Mammals |
| **LENGTH/WEIGHT** | up to 10 ft. including tail/620 lb. |
| **DIET** | herbivore: grass, plants, leaves |
| **LOCATION** | southern and eastern Africa |
| **STATUS** | Least Concern |

## • January 18th •
## HOOPOE

This exotic bird has startling black-and-white stripes on its wings and a distinctive crest of feathers that it whisks up when it is excited or coming in to land. It uses its long, curved beak to forage in the grass for bugs and worms. Its name comes from its soft repeated call, and when searching for a mate, males take part in song duels to put off any rivals.

| | |
|---|---|
| **SCIENTIFIC NAME** | *Upapa epops* |
| **ANIMAL GROUP** | Birds |
| **WINGSPAN/WEIGHT** | 18 in./3 oz. |
| **DIET** | omnivore: insects, worms, spiders, berries, seeds |
| **LOCATION** | Europe, Asia, Africa |
| **STATUS** | Least Concern |

## • January 19th •
## ARMADILLO GIRDLED LIZARD

As well as being covered in spines and scales, this lizard has another trick up its sleeve when threatened. It curls into a ball with its tail in its mouth, leaving a predator such as a bird of prey puzzling how to continue its attack. The lizard lives in sandstone crevices on mountain slopes.

| | |
|---|---|
| **SCIENTIFIC NAME** | *Ouroborus cataphractus* |
| **ANIMAL GROUP** | Reptiles |
| **LENGTH/WEIGHT** | up to 8 in. including tail/17 lb. |
| **DIET** | omnivore: insects, spiders, scorpions, plants |
| **LOCATION** | western South Africa |
| **STATUS** | Near Threatened |

• January 20th •
# LONGHORN COWFISH

Also called the horned boxfish because of its shape, this fish exudes a toxin through its skin if attacked. It protects the reefs it lives in by eating invertebrates that would destroy the corals.

| | |
|---|---|
| **SCIENTIFIC NAME** | *Lactoria cornuta* |
| **ANIMAL GROUP** | Fish |
| **LENGTH/WEIGHT** | up to 18 in./5 oz. |
| **DIET** | omnivore: sponges, algae, mollusks, crustaceans, worms |
| **LOCATION** | Indo-Pacific |
| **STATUS** | Not Evaluated |

• January 21st •
# CHINESE WATER DEER

A small Asian deer native to China and Korea, this "water deer" lives in the reeds and rushes alongside rivers, as well as in swampy grasslands. It is an excellent swimmer. The buck does not have antlers, but it uses its small tusks as weapons to fight off other males and to protect itself. This is a solitary animal, but it will give a barking warning of danger approaching for other deer.

| | |
|---|---|
| **SCIENTIFIC NAME** | *Hydropotes inermis* |
| **ANIMAL GROUP** | Mammals |
| **LENGTH/WEIGHT** | up to 3 ft./40 lb. |
| **DIET** | herbivore: weeds, grass, leaves |
| **LOCATION** | Asia |
| **STATUS** | Vulnerable |

• January 22nd •
# EASTERN CORAL SNAKE

Very secretive, this snake spends most of its time sheltering under rocks or in burrows. It hunts on the ground in the daytime, and its extremely venomous bite causes instant paralysis in prey. When its babies hatch, they are only 7 inches long, but already fully venomous and able to hunt.

| | |
|---|---|
| **SCIENTIFIC NAME** | *Micrurus fulvius* |
| **ANIMAL GROUP** | Reptiles |
| **LENGTH/WEIGHT** | up to 4 ft./5 lb. |
| **DIET** | carnivore: lizards, other snakes, birds, frogs, fish, insects |
| **LOCATION** | southeastern North America |
| **STATUS** | Least Concern |

• January 23rd •

# MOOSE

Known as moose in North America and elk in Europe and Asia, these are the world's largest and heaviest deer. The bulls (males) have antlers spreading up to 7 feet, which they use to impress females and fight off rival males, as well as, occasionally, see off a predator. They shed the 60-pound antlers in winter, allowing the bulls to store more energy. Each spring the antlers are regrown.

| | |
|---|---|
| **SCIENTIFIC NAME** | *Alces alces* |
| **ANIMAL GROUP** | Mammals |
| **LENGTH/WEIGHT** | up to 10 ft. including tail/1,550 lb. |
| **DIET** | herbivore: aquatic plants, grass, shrubs, leaves |
| **LOCATION** | northern North America, Europe and Asia |
| **STATUS** | Least Concern |

• January 24th •
# INDIAN FLYING FOX

The reason this "fox" can fly is that it is in fact a fox-faced bat! It is one of the largest bats in the world. By day, it rests hanging upside down with hundreds of others at a roosting site. By night, it takes off in search of food. These bats travel an average of 19 miles a night looking for ripe fruit such as figs, mangoes and bananas, as well as nectar from flowers, although they may travel much further. They drink water on the wing from rivers, and also from leaves after rain.

| | |
|---|---|
| **SCIENTIFIC NAME** | *Pteropus medius* |
| **ANIMAL GROUP** | Mammals |
| **WINGSPAN/WEIGHT** | up to 5 ft./4.4 lb. |
| **DIET** | herbivore: fruit, blossoms, nectar |
| **LOCATION** | southern Asia |
| **STATUS** | Least Concern |

• January 25th •
# AFRICAN FISH EAGLE

Swooping in with talons outstretched, this bird of prey, also known as the "screaming eagle," snatches a fish from the surface of the water. It had sighted its prey from above and will carry it away to a perch or its nest to eat. As well as fish, it hunts ducks, terrapins and even flamingos (see p.37). It also steals food from other birds such as herons (see p.98).

| | |
|---|---|
| **SCIENTIFIC NAME** | *Haliaeetus vocifer* |
| **ANIMAL GROUP** | Birds |
| **WINGSPAN/WEIGHT** | up to 8 ft./8 lb. |
| **DIET** | carnivore: mainly fish, other waterfowl |
| **LOCATION** | Africa, south of the Sahara |
| **STATUS** | Least Concern |

# RECORD BREAKERS

The world is full of extraordinary wild animals that have developed breathtaking skills, whether to find food, find a mate or simply to survive. Some animals hold world records for their particular abilities. Here are some of these high achievers.

• January 26th •

## ARCTIC TERN

This small bird holds the record for the longest migration of any bird. Every year, it travels from the Arctic Circle, where it breeds, south to the Antarctic to feed, and back again – a round trip of 50,000 miles.

| | |
|---|---|
| **SCIENTIFIC NAME** | *Sterna paradisaea* |
| **ANIMAL GROUP** | Birds |
| **WINGSPAN/WEIGHT** | up to 2.8 ft./4.2 oz. |
| **DIET** | carnivore: mainly fish, also crustaceans, insects |
| **LOCATION** | Arctic and Antarctica via Europe and Africa |
| **STATUS** | Least Concern |

• January 27th •

## LAYSAN ALBATROSS

A large seabird that glides over the ocean for hundreds of miles a day with hardly a wingbeat, this albatross is known to have the longest life of any bird. These birds breed every year, and one has even been recorded laying an egg at the age of 70.

| | |
|---|---|
| **SCIENTIFIC NAME** | *Phoebastria immutabilis* |
| **ANIMAL GROUP** | Birds |
| **WINGSPAN/WEIGHT** | up to 7 ft./9 lb. |
| **DIET** | carnivore: mainly squid, fish |
| **LOCATION** | northern Pacific Ocean |
| **STATUS** | Near Threatened |

• January 28th •

## INDO-PACIFIC SAILFISH

The fastest species of fish over short distances, the sailfish has been recorded traveling at 68 mph. It can pull down its large dorsal sail fin and pull in the two pectoral fins to streamline its body. This allows it to move faster through the water in pursuit of a school of prey such as sardines.

| | |
|---|---|
| **SCIENTIFIC NAME** | *Istiophorus platypterus* |
| **ANIMAL GROUP** | Fish |
| **LENGTH/WEIGHT** | up to 11 ft./220 lb. |
| **DIET** | carnivore: cephalopods, fish |
| **LOCATION** | Indian, Pacific and Atlantic Oceans |
| **STATUS** | Vulnerable |

• January 29th •

## KOALA

Napping for up to 18 hours out of every 24, the koala is the sleepiest marsupial. It lives on a diet of eucalyptus leaves, eating no more than 1.8 pounds a day. This gives it very little energy, so it needs to conserve what it has by sleeping.

| | |
|---|---|
| SCIENTIFIC NAME | *Phascolarctos cinereus* |
| ANIMAL GROUP | Mammals |
| LENGTH/WEIGHT | up to 2.8 ft./26 lb. |
| DIET | herbivore: mainly eucalyptus leaves, also paperbark and brush box trees |
| LOCATION | Australia |
| STATUS | Vulnerable |

• January 30th •

## CHEETAH

With four 22-foot-long strides a second, a slender body and long legs built for speed, this animal can move! It is the fastest mammal on land over short distances with a record-breaking speed of 65 mph.

| | |
|---|---|
| SCIENTIFIC NAME | *Acinonyx jubatus* |
| ANIMAL GROUP | Mammals |
| LENGTH/WEIGHT | up to 7 ft. including tail/145 lb. |
| DIET | carnivore: antelope, warthogs, oryx, game birds, rabbits |
| LOCATION | Africa |
| STATUS | Vulnerable |

• January 31st •

## CHINESE GIANT SALAMANDER

A living fossil with relatives that date back to the time of the dinosaurs, this is the world's largest amphibian. It lives underwater in fast-flowing rivers, but does not have gills. Instead, it absorbs oxygen from the water through its skin.

| | |
|---|---|
| SCIENTIFIC NAME | *Andrias davidianus* |
| ANIMAL GROUP | Amphibians |
| LENGTH/WEIGHT | up to 6 ft. including tail/110 lb. |
| DIET | carnivore: worms, insects, fish, frogs |
| LOCATION | eastern Asia |
| STATUS | Critically Endangered |

# FEBRUARY

• February 1st •

## SALTWATER CROCODILE

This is a dangerous and well-equipped killer. Curved, conical teeth, some more than 4 inches long, line its jaws – jaws that have the strongest bite of any animal. It can sprint short distances on land and in the water it is a powerful swimmer. "Salties" are found in freshwater rivers, estuaries and mangrove swamps, as well as the open ocean. They float near the shore with only their eyes showing, patiently waiting for prey as large as water buffalo to come to drink, before grabbing them.

| | |
|---|---|
| **SCIENTIFIC NAME** | *Crocodylus porosus* |
| **ANIMAL GROUP** | Reptiles |
| **LENGTH/WEIGHT** | up to 23 ft./1.3 tons |
| **DIET** | carnivore: water buffalo, monkeys, wild boar, kangaroos |
| **LOCATION** | Australia, southern and southeastern Asia |
| **STATUS** | Least Concern |

• February 2nd •

## SNAIL KITE

This bird of prey has a healthy appetite for large apple snails. It snatches up these tasty aquatic morsels when they surface to breathe in rivers and marshes. Holding its prey in one foot, it returns to its perch, where it extracts the snail from its shell with a hooked beak specially shaped for the task.

| | |
|---|---|
| SCIENTIFIC NAME | *Rostrhamus sociabilis* |
| ANIMAL GROUP | Birds |
| WINGSPAN/WEIGHT | up to 4 ft./1.3 lb. |
| DIET | carnivore: mainly large freshwater snails |
| LOCATION | South and Central America, southern North America |
| STATUS | Least Concern |

• February 3rd •

## LEAST CHIPMUNK

Easily recognized by a striped back and a tail that is as long as its body, this smallest member of its family is aptly named. It has pouches in its cheeks that it stuffs with food to carry back to its burrow and store for winter. When it gets cold, these ground squirrels go into a state of semi-hibernation called "torpor." They wake from time to time and eat from the stored food.

| | |
|---|---|
| SCIENTIFIC NAME | *Neotamias minimus* |
| ANIMAL GROUP | Mammals |
| LENGTH/WEIGHT | up to 9 in. including tail/2 oz. |
| DIET | herbivore: seeds, nuts, fruits, berries |
| LOCATION | North America |
| STATUS | Least Concern |

• February 4th •

# HERMANN'S TORTOISE

Like all tortoises, this one has a bony shell. The domed top is called the "carapace" and the flat layer underneath is the "plastron." The carapace is covered by a layer of linked pieces called "scutes," which are made of keratin, the material that forms human hair and nails. Active during the day, the tortoise uses its sense of smell to find food and shelters when it gets hot. It hibernates (see pp.174-175) in winter.

| | |
|---|---|
| **SCIENTIFIC NAME** | Testudo hermanni |
| **ANIMAL GROUP** | Reptiles |
| **LENGTH/WEIGHT** | up to 11 in./9 lb. |
| **DIET** | herbivore: mainly fruit, herbs, berries, flowers, leaves |
| **LOCATION** | southern Europe, western Asia |
| **STATUS** | Near Threatened |

• February 5th •

# LEAFY SEADRAGON

With leaflike fins and frilly attachments to its head and body, this bony fish could not be better camouflaged in beds of kelp. It moves upright through the water using its fins to propel itself along. It does not have a stomach, so has to eat almost constantly to survive, sucking up food through its tube-shaped mouth like a straw.

| | |
|---|---|
| **SCIENTIFIC NAME** | Phycodurus eques |
| **ANIMAL GROUP** | Fish |
| **LENGTH/WEIGHT** | up to 14 in./4 oz. |
| **DIET** | omnivore: mainly shrimp, also plankton and larvae |
| **LOCATION** | southern and western coasts of Australia |
| **STATUS** | Least Concern |

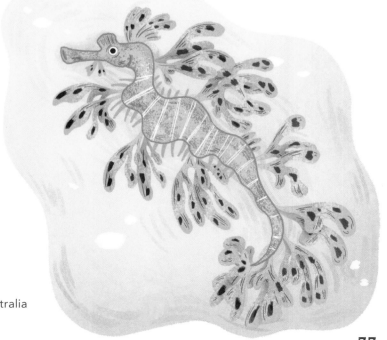

• February 6th •

## BLUE POISON FROG

Like most of the other poison frogs, the bright color of this hunchbacked amphibian acts as a warning. If this does not work, there are toxins in its skin that can paralyze or kill a predator. These frogs live only in the rainforests of Suriname and northern Brazil, under rocks and moss near streams and also high up in the trees.

| | |
|---|---|
| **SCIENTIFIC NAME** | *Dendrobates tinctorius* |
| **ANIMAL GROUP** | Amphibians |
| **LENGTH/WEIGHT** | up to 1.8 in./0.3 oz. |
| **DIET** | carnivore: ants, termites, beetles |
| **LOCATION** | South America |
| **STATUS** | Least Concern |

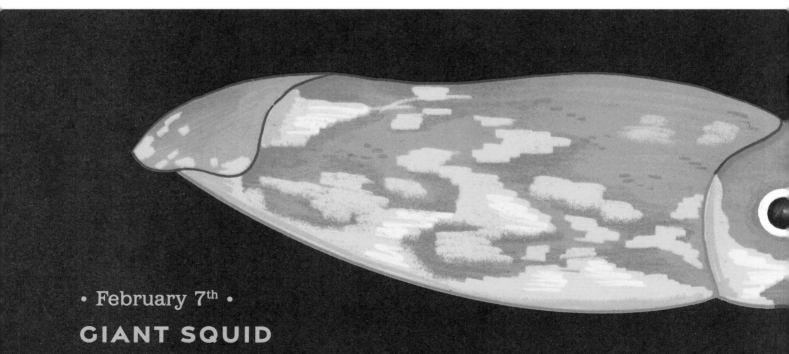

• February 7th •

## GIANT SQUID

Little is known about this extraordinary creature as few people have actually seen one. It is the largest cephalopod (a mollusk that has its tentacles attached to its head) in the world. Its enormous eyes are around 12 inches across and it has a sharp beak in the centre of its eight thick arms that slices prey into bite-sized pieces.

| | |
|---|---|
| **SCIENTIFIC NAME** | *Architeuthis dux* |
| **ANIMAL GROUP** | Invertebrates |
| **LENGTH/WEIGHT** | up to 66 ft./1.1 tons |
| **DIET** | carnivore: fish and other squid |
| **LOCATION** | oceans worldwide |
| **STATUS** | Least Concern |

• February 8th •

# JEWEL BEETLE

This beautiful wood-boring beetle lays its eggs in the dead branches of decaying pine trees, high up where the sun can warm the wood. It belongs to a family known by the name "jewel beetle" because of their shiny iridescent colors. They are among the largest of all the beetles. Unfortunately this particular one is threatened with extinction in many parts of Europe because of the destruction of the forests where it lives.

| | |
|---|---|
| **SCIENTIFIC NAME** | *Buprestis splendens* |
| **ANIMAL GROUP** | Invertebrates |
| **LENGTH** | up to 0.9 in. |
| **DIET** | herbivore: wood bark, leaves, flowers |
| **LOCATION** | western Europe |
| **STATUS** | Endangered |

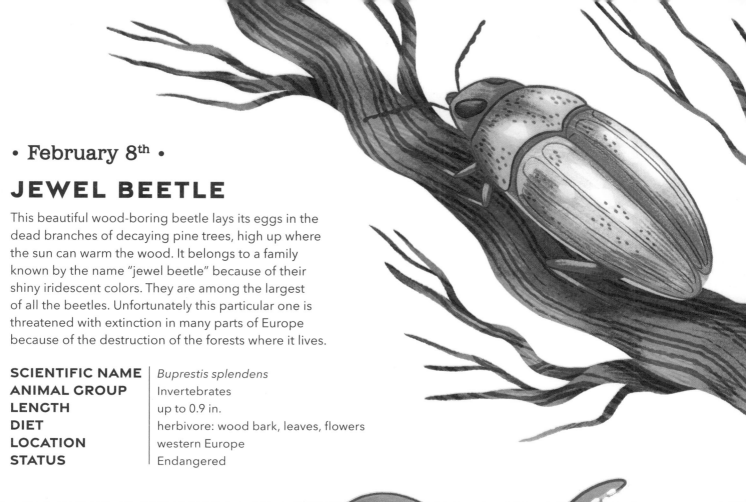

# LIVING WITH A VOLCANO

Some animals live in the most extreme of Earth's environments, near or even on active volcanoes. The animals on these two pages live successfully in habitats that may be destroyed at any moment if there is an eruption.

• February 9th •

## TARDIGRADE

This is the ultimate animal survivor. Also known as a water bear, this eight-legged microanimal is really tough and adaptable. It can survive not only in active lava fields but also on Antarctic glaciers and in space!

| | |
|---|---|
| SCIENTIFIC NAME | *Macrobiotus sapiens* |
| ANIMAL GROUP | Invertebrates |
| LENGTH | up to 0.05 in. |
| DIET | omnivore: plant cells, algae, small invertebrates |
| LOCATION | worldwide |
| STATUS | Not Evaluated |

*As seen under a microscope*

• February 10th •

## HAWAIIAN HOARY BAT

This solitary nocturnal hunter is found on all of the main volcanic Hawaiian Islands. During the day, it roosts in trees, caves or lava tubes near the mountain summits.

| | |
|---|---|
| SCIENTIFIC NAME | *Lasiurus semotus* |
| ANIMAL GROUP | Mammals |
| WINGSPAN/WEIGHT | up to 17 in./0.6 oz. |
| DIET | carnivore: moths, beetles, crickets, termites |
| LOCATION | Hawaiian Islands |
| STATUS | Not Evaluated |

• February 11th •

## POMPEII WORM

This worm lives in tubes anchored to hydrothermal smokers deep in the ocean. These vents heat saltwater with magma and the worm can tolerate up to 176°F!

| | |
|---|---|
| SCIENTIFIC NAME | *Alvinella pompejana* |
| ANIMAL GROUP | Invertebrates |
| LENGTH | up to 6 in. |
| DIET | omnivore: bacteria |
| LOCATION | Pacific Ocean |
| STATUS | Not Evaluated |

• February 12th •

# GALÁPAGOS PINK LAND IGUANA

There are only 200 of these iguanas in the world, and they all live on the island of Isabela at the foot of a volcano that is constantly active, and last erupted in 2015. Like all reptiles, they are cold-blooded and need to be near a source of heat, so they bask on the volcanic rocks in the sun.

| | |
|---|---|
| SCIENTIFIC NAME | *Conolophus marthae* |
| ANIMAL GROUP | Reptiles |
| LENGTH/WEIGHT | up to 3 ft./15 lb. |
| DIET | herbivore: plants, grasses, cacti |
| LOCATION | Galápagos Islands |
| STATUS | Critically Endangered |

• February 13th •

# LESSER FLAMINGO

Lake Natron in Tanzania is near one of Africa's active volcanoes, and its salt-filled waters are deadly to wildlife. Yet 75 percent of all lesser flamingos live there. Special tough skin and the scales on their legs prevent burns. And they can drink water near the boiling point and eat toxic algae because their nasal cavities filter out salt.

| | |
|---|---|
| SCIENTIFIC NAME | *Phoeniconaias minor* |
| ANIMAL GROUP | Birds |
| WINGSPAN/WEIGHT | up to 4 ft./4 lb. |
| DIET | omnivore: blue-green algae, diatoms, rotifers |
| LOCATION | eastern and southern Africa, Asia |
| STATUS | Near Threatened |

• February 14th •

# VAMPIRE GROUND FINCH

Home for this unusual species is the remote volcanic Wolf and Darwin Islands, where food and water are often difficult to find. If there are seeds and insects it eats those, but if not, it drinks the blood of blue-footed boobies (see p. 107).

| | |
|---|---|
| SCIENTIFIC NAME | *Geospiza septentrionalis* |
| ANIMAL GROUP | Birds |
| WINGSPAN/WEIGHT | up to 9 in./0.7 oz |
| DIET | omnivore: seeds, insects, blood |
| LOCATION | Galápagos Islands |
| STATUS | Vulnerable |

• **February 15th** •

# EMPEROR PENGUIN

For four months on the frozen ocean of Antarctica, male emperor penguins protect their eggs and young from temperatures as low as minus 58°F and winds that travel at up to 125 mph. They each look after a single egg until it hatches, then huddle together in a big group, taking turns to be on the outside in the bitter winds. They do this until the females return from the sea to take over feeding the chicks and keeping them warm.

| | |
|---|---|
| **SCIENTIFIC NAME** | *Aptenodytes forsteri* |
| **ANIMAL GROUP** | Birds |
| **HEIGHT/WEIGHT** | up to 4 ft./90 lb. |
| **DIET** | carnivore: fish, krill, squid |
| **LOCATION** | Antarctica |
| **STATUS** | Near Threatened |

• February 16th •

# CRESTED PORCUPINE

Covered in sharp quills to deter predators, this is the world's largest rodent. It relies on its sense of hearing and smell because it has poor eyesight. If threatened, it chatters its teeth and produces a smell before turning to raise its quills and make itself look larger. It will then run backward and ram into the predator, injuring it with the quills.

| | |
|---|---|
| **SCIENTIFIC NAME** | *Hystrix cristata* |
| **ANIMAL GROUP** | Mammals |
| **LENGTH/WEIGHT** | up to 3.6 ft. including tail/65 lb. |
| **DIET** | omnivore: bark, roots, bulbs, insects |
| **LOCATION** | northern Africa, southern Europe |
| **STATUS** | Least Concern |

• February 17th •

# OCELOT

Although it is twice the size of one, this animal looks much like a domestic cat. It is an excellent climber and swimmer. Mostly active at night, it uses its keen hearing and sharp sight to hunt all kinds of prey in the jungles and forests where it lives. It does not chew its food. Instead it tears the meat into pieces which it swallows whole.

| | |
|---|---|
| **SCIENTIFIC NAME** | *Leopardus pardalis* |
| **ANIMAL GROUP** | Mammals |
| **LENGTH/WEIGHT** | up to 5 ft. including tail/35 lb. |
| **DIET** | carnivore: rabbits, rodents, iguanas, fish, frogs |
| **LOCATION** | southern North America, Central and South America |
| **STATUS** | Least Concern |

• February 18th •

# VIETNAMESE GIANT CENTIPEDE

Large and traveling very fast across the forest floor, this tropical centipede has brightly colored legs at the end of which are sharp claws. It is an aggressive hunter that uses its two front claws to subdue and inject venom into prey, or to defend itself against a predator. During the day, it rests under stones, rotting wood or loose bark.

| | |
|---|---|
| **SCIENTIFIC NAME** | Scolopendra dehaani |
| **ANIMAL GROUP** | Invertebrates |
| **LENGTH** | up to 8 in. |
| **DIET** | carnivore: insects, spiders, other centipedes |
| **LOCATION** | southern and eastern Asia |
| **STATUS** | Not Evaluated |

• February 19th •

# ELF OWL

This tiny, short-tailed owl lives in dry or desert regions. It is the smallest owl in the Sonoran Desert, where it often nests in holes in saguaro cacti 33 feet above the ground. Like other owls, it hunts at night, and although it mainly eats insects, it will also take scorpions and small lizards. If it is threatened or even caught, it plays dead. This encourages a predator such as another owl, snake or bobcat to ignore it or relax its grip so that the elf owl can escape. If an intruder gets near its nest, it may also make a loud barking sound and clap its bill to put them off.

| | |
|---|---|
| **SCIENTIFIC NAME** | Micrathene whitneyi |
| **ANIMAL GROUP** | Birds |
| **WINGSPAN/WEIGHT** | up to 13 in./1.8 oz. |
| **DIET** | carnivore: moths, crickets, scorpions, beetles |
| **LOCATION** | southern North America |
| **STATUS** | Least Concern |

# ANIMALS AROUND THE HOUSE

Have you ever thought about the animals (not the human ones) with which you share your house? Maybe some of these live with you undetected, or visit you once a year or even more regularly. Try listening hard – you may hear them moving around.

• February 20th •

## BARN SWALLOW

Swallows migrate (*see pp.84–85*) to the same place every year to raise their chicks. Many build their cup-shaped nests of mud and straw under the eaves of houses or in barns.

| | |
|---|---|
| SCIENTIFIC NAME | *Hirundo rustica* |
| ANIMAL GROUP | Birds |
| WINGSPAN/WEIGHT | up to 14 in./0.9 oz. |
| DIET | carnivore: flying insects, beetles |
| LOCATION | worldwide except Antarctica |
| STATUS | Least Concern |

• February 21st •

## BROWN RECLUSE SPIDER

Also known as the violin spider (it has a violin-shaped patch on its head), this little housemate has a venomous bite. It likes warm places such as inside drawers or behind furniture.

| | |
|---|---|
| SCIENTIFIC NAME | *Loxosceles reclusa* |
| ANIMAL GROUP | Invertebrates |
| LEGSPAN/WEIGHT | up to 1.5 in./0.2 oz. |
| DIET | carnivore: small insects, other spiders |
| LOCATION | North America |
| STATUS | Not Evaluated |

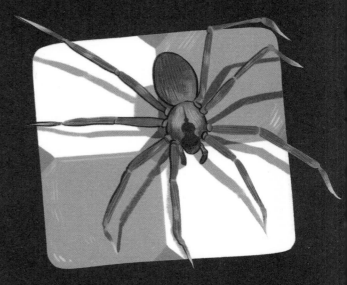

• February 22nd •

## NORTHERN RACCOON

This solitary nocturnal mammal likes city life, gardens and parks. If you hear thumps above your head, there may be a raccoon either on the roof or breaking into the attic to nest.

| | |
|---|---|
| SCIENTIFIC NAME | *Procyon lotor* |
| ANIMAL GROUP | Mammals |
| LENGTH/WEIGHT | up to 3.3 ft. including tail/26 lb. |
| DIET | omnivore: fruit, nuts, eggs and baby birds, frogs, worms, garbage |
| LOCATION | North America |
| STATUS | Least Concern |

• February 23rd •
## COMMON HOUSE GECKO

Geckos are attracted by electric lights because they hunt the insects they find around them. A gecko can scuttle up walls and upside down across ceilings because tiny, sticky hairs called "setae" cover the base of its toes. These allow the gecko to cling to surfaces.

| | |
|---|---|
| **SCIENTIFIC NAME** | *Hemidactylus frenatus* |
| **ANIMAL GROUP** | Reptiles |
| **LENGTH/WEIGHT** | up to 6 in./2.5 oz. |
| **DIET** | carnivore: crickets, fruit flies, silkworms |
| **LOCATION** | southeastern Asia |
| **STATUS** | Least Concern |

• February 24th •
## HOUSE MOUSE

This is one of the most common of all the rodents, and most houses have mice living in them at one time or another. They can have up to 14 litters of young a year with 12 mice in each litter. That's a lot of house mice!

| | |
|---|---|
| **SCIENTIFIC NAME** | *Mus musculus* |
| **ANIMAL GROUP** | Mammals |
| **LENGTH/WEIGHT** | up to 8 in. including tail/1.1 oz. |
| **DIET** | omnivore: seeds, leaves, insects, carrion |
| **LOCATION** | worldwide except Antarctica |
| **STATUS** | Least Concern |

• February 25th •
## COMMON PIPISTRELLE

In summer these small bats roost during the day and hunt insects at dusk. In winter, they hibernate (see pp. 174-175) in buildings, sometimes in small colonies in the roof spaces where it is warm.

| | |
|---|---|
| **SCIENTIFIC NAME** | *Pipistrellus pipistrellus* |
| **ANIMAL GROUP** | Mammals |
| **WINGSPAN/WEIGHT** | up to 9 in./0.3 oz. |
| **DIET** | carnivore: flies, lacewings, mosquitoes |
| **LOCATION** | Europe, northern Africa, southern Asia |
| **STATUS** | Least Concern |

• February 26th •

# BLACK MAMBA

Fast, aggressive and lethal, this is a very dangerous snake and one of the world's most venomous. It is called "black" because of the inside of its mouth, which is shown when it is threatened. The rest of the snake can vary in color. It can travel at up to 12 mph and two drops of its venom can kill a human.

| | |
|---|---|
| **SCIENTIFIC NAME** | *Dendroaspis polylepis* |
| **ANIMAL GROUP** | Reptiles |
| **LENGTH/WEIGHT** | up to 14 ft./3.5 lb. |
| **DIET** | carnivore: small mammals and birds |
| **LOCATION** | Africa south of the Sahara |
| **STATUS** | Least Concern |

• February 27th •

# PRONGHORN

The black, forked horns of the male curve inward and are unique to this family of deerlike mammals. The horns can grow up to 20 inches in length, while those of the female are shorter and usually straight. The pronghorn is the second-fastest animal on land after the cheetah (*see p.29*), able to reach a speed of up to 60 mph when startled by predators such as wolves or cougars.

| | |
|---|---|
| **SCIENTIFIC NAME** | *Antilocapra americana* |
| **ANIMAL GROUP** | Mammals |
| **LENGTH/WEIGHT** | up to 5 ft./145 lb. |
| **DIET** | herbivore: grasses, shrubs, flowers, fruit |
| **LOCATION** | North America |
| **STATUS** | Least Concern |

• February 28th •

# BLACK-LEGGED KITTIWAKE

Perched on a ledge high on a clifftop, these medium-sized gulls are out of reach of most predators. The female usually lays one or two eggs, and both parents incubate the eggs and share feeding duties. Kittiwakes hunt in flocks through the spring, summer and autumn, then spend the winter months out at sea.

| | |
|---|---|
| **SCIENTIFIC NAME** | *Rissa tridactyla* |
| **ANIMAL GROUP** | Birds |
| **WINGSPAN/WEIGHT** | up to 3.6 ft./1.1 lb. |
| **DIET** | carnivore: fish, shrimps and worms |
| **LOCATION** | northern Pacific and Atlantic Oceans, Arctic Ocean |
| **STATUS** | Vulnerable |

• February 29th •

# PYGMY HIPPOPOTAMUS

Secretive and mostly nocturnal, this small hippo stays hidden in swamps or in hollows under the banks of forest streams where it lives. It is semiaquatic, spending time both on land and in the water, and finding its food in both places. Its skin contains pores that secrete a thick, oily substance that acts as a moisturizer and water repellent. This allows it to remain in water for long periods of time.

| | |
|---|---|
| **SCIENTIFIC NAME** | *Choeropsis liberiensis* |
| **ANIMAL GROUP** | Mammals |
| **LENGTH/WEIGHT** | up to 6 ft. including tail/600 lb. |
| **DIET** | herbivore: ferns, grasses, fruit, aquatic plants |
| **LOCATION** | western Africa |
| **STATUS** | Endangered |

# MARCH

• March 1st •

## MANED WOLF

Standing nearly 3 feet tall at the shoulder, this large member of the dog family has very long legs. It often hunts in tall grass, and its height allows it to easily spot small prey. A female may give birth to up to five pups at a time. She nurses them and teaches them to hunt before they become independent around the age of one.

| | |
|---|---|
| SCIENTIFIC NAME | *Chrysocyon brachyurus* |
| ANIMAL GROUP | Mammals |
| LENGTH/WEIGHT | up to 6 ft. including tail/50 lb. |
| DIET | omnivore: rabbits, rodents, insects, fruit, plants |
| LOCATION | central and eastern South America |
| STATUS | Near Threatened |

• March 2nd •

# KING COBRA

This is the longest of all the venomous snakes and it is very dangerous. It is referred to as "king" because it can kill and eat other cobras. It is fierce and defensive. When confronted by danger, it holds its ground, raising the front part of its body around 4 feet into the air, while hissing and hooding to make itself look bigger. One bite from its fangs can inject enough venom to kill an elephant.

| | |
|---|---|
| **SCIENTIFIC NAME** | *Ophiophagus hannah* |
| **ANIMAL GROUP** | Reptiles |
| **LENGTH/WEIGHT** | up to 18 ft./21 lb. |
| **DIET** | carnivore: other snakes, birds, lizards, rodents |
| **LOCATION** | Asia |
| **STATUS** | Vulnerable |

• March 3rd •

# RED-EYED TREE FROG

These tiny frogs have a great way to camouflage themselves when they are sleeping in the rainforest trees. They tuck their legs under their bodies and close their eyes so they appear to be all green. Suction cups on their toes keep them attached to the branches.

| | |
|---|---|
| **SCIENTIFIC NAME** | *Agalychnis callidryas* |
| **ANIMAL GROUP** | Amphibians |
| **LENGTH/WEIGHT** | up to 3.1 in./0.5 oz. |
| **DIET** | carnivore: crickets, amphibians, fruit flies, moths |
| **LOCATION** | Central America, northern South America |
| **STATUS** | Least Concern |

• March 4th •

# QUEEN ALEXANDRA'S BIRDWING

Only found in a small area of rainforest on the island of Papua New Guinea, this swallowtail butterfly develops from a large black caterpillar with red spikes. Both adult and young feed only on a couple of species of vine, which are poisonous to most animals. Any hungry predator will find that the caterpillar is poisonous too!

| | |
|---|---|
| **SCIENTIFIC NAME** | *Ornithoptera alexandrae* |
| **ANIMAL GROUP** | Invertebrates |
| **WINGSPAN/WEIGHT** | up to 12 in./0.4 lb. |
| **DIET** | herbivore: nectar |
| **LOCATION** | Asia – Papua New Guinea |
| **STATUS** | Endangered |

# IN THE RAINFOREST

Tropical rainforests are hot, humid and wet. Trees grow tall to reach the sunlight. Animals that live here are adapted to live at different levels, some on the dark forest floor, others in the understory, most in the canopy and birds high up at the very top.

• March 5th •

## RETICULATED PYTHON

This snake lives in the rainforest canopy where it waits in hiding to trap and constrict passing prey. It is never very far from ponds, streams or rivers and swims very well.

| | |
|---|---|
| **SCIENTIFIC NAME** | *Malayopython reticulatus* |
| **ANIMAL GROUP** | Reptiles |
| **LENGTH/WEIGHT** | up to 30 ft./350 lb. |
| **DIET** | carnivore: small mammals, pigs, civets, bearcats, monkeys, small deer |
| **LOCATION** | southern and southeastern Asia |
| **STATUS** | Least Concern |

• March 6th •

## RED-BELLIED PIRANHA

Large shoals of these aggressive predators swim in rainforest lakes and rivers. They can kill large prey such as capybaras, rapidly stripping flesh from bone.

| | |
|---|---|
| **SCIENTIFIC NAME** | *Pygocentrus nattereri* |
| **ANIMAL GROUP** | Fish |
| **LENGTH/WEIGHT** | up to 14 in./4 lb. |
| **DIET** | omnivore: insects, crustaceans, worms, fish, capybaras, fruit, seeds |
| **LOCATION** | South America |
| **STATUS** | Not Evaluated |

• March 7th •

## TIGER QUOLL

Very agile both on the forest floor and in the trees, this marsupial travels up to 2.5 miles a night to hunt for food. It shelters in burrows, tree holes or rocky crevices by day.

| | |
|---|---|
| **SCIENTIFIC NAME** | *Dasyurus maculatus* |
| **ANIMAL GROUP** | Mammals |
| **LENGTH/WEIGHT** | up to 3.6 ft. including tail/15 lb. |
| **DIET** | carnivore: possums, bandicoots, rats, reptiles, birds, insects |
| **LOCATION** | Australia |
| **STATUS** | Near Threatened |

• March 8th •

# CRESTED SERPENT EAGLE

This raptor builds its large platform nest high up, where it has a clear view of the surrounding area. It blends well into the rainforest foliage of the canopy, which makes it a very effective hunter of snakes and other prey.

| | |
|---|---|
| SCIENTIFIC NAME | *Spilornis cheela* |
| ANIMAL GROUP | Birds |
| WINGSPAN/WEIGHT | up to 5 ft./4 lb. |
| DIET | carnivore: snakes, small mammals, monkeys, birds |
| LOCATION | Asia |
| STATUS | Least Concern |

• March 9th •

# BONGO

Only found in rainforests with dense undergrowth, this antelope has to hide from leopards, hyenas and lions. It feeds mainly at night, using its long tongue to pull up roots.

| | |
|---|---|
| SCIENTIFIC NAME | *Tragelaphus eurycerus* |
| ANIMAL GROUP | Mammals |
| LENGTH/WEIGHT | up to 8 ft. including tail/880 lb. |
| DIET | herbivore: leaves, bark, grasses, roots |
| LOCATION | western Africa |
| STATUS | Near Threatened |

• March 10th •

# BONOBO

These great apes live in groups led by a female. During the day, they forage for food both on the ground and in the canopy. At night, they sleep in nests in the forks of trees.

| | |
|---|---|
| SCIENTIFIC NAME | *Pan paniscus* |
| ANIMAL GROUP | Mammals |
| HEIGHT/WEIGHT | up to 4.3 ft./90 lb. |
| DIET | omnivore: fruit, insects, fish, small mammals, worms |
| LOCATION | central Africa |
| STATUS | Endangered |

• March 11th •

# TAWNY FROGMOUTH

Although this bird looks like an owl, it belongs to a different family altogether, that of the nightjar. A tough, nocturnal bird, it lives in different kinds of habitats from deserts to cold mountains. The owl's thick feathers insulate it well, both from cold and heat. When it roosts in the trees during the day, it is well camouflaged, stretching its neck out so it looks just like a dead branch.

| | |
|---|---|
| **SCIENTIFIC NAME** | *Podargus strigoides* |
| **ANIMAL GROUP** | Birds |
| **WINGSPAN/WEIGHT** | up to 40 in./1.5 lb. |
| **DIET** | carnivore: insects, spiders, worms, cockroaches, snails, small mammals and reptiles, frogs |
| **LOCATION** | Australia |
| **STATUS** | Least Concern |

• March 12th •

# RIBBON EEL

All ribbon eels are hermaphrodites. They begin their lives as males and are black and yellow. As they age, the males turn bright blue and yellow. At the final stage they become all-yellow females and are able to lay eggs. Also known as leaf-nosed moray eels, these are aggressive hunters.

| | |
|---|---|
| **SCIENTIFIC NAME** | *Rhinomuraena quaesita* |
| **ANIMAL GROUP** | Fish |
| **LENGTH** | up to 4.3 ft. |
| **DIET** | carnivore: shrimps, small fish |
| **LOCATION** | Indian and Pacific Oceans |
| **STATUS** | Least Concern |

• March 13th •

# RING-TAILED LEMUR

Only found in the southern forests and shrublands of the island of Madagascar, "mobs," or groups, of up to 25 lemurs can be seen sunbathing with their bellies upturned and arms and legs stretched out. When they travel through the forest, they raise their tails to keep the group together.

| | |
|---|---|
| **SCIENTIFIC NAME** | *Lemur catta* |
| **ANIMAL GROUP** | Mammals |
| **LENGTH/WEIGHT** | up to 3.6 ft. including tail/7 lb. |
| **DIET** | omnivore: leaves, flowers, insects, fruit |
| **LOCATION** | Africa |
| **STATUS** | Endangered |

• March 14th •

# SPECTACLED CAIMAN

Most caimans live in freshwater rivers and streams, but this one also tolerates saltwater, which gives it a larger range. It mainly hunts at night, staying submerged during the day, and has a reputation for eating anything. It may even have over 100 different animals as prey! It burrows into the mud in summer and estivates, a form of hibernation (see pp. 174–175), to avoid extreme heat.

| | |
|---|---|
| **SCIENTIFIC NAME** | *Caiman crocodilus* |
| **ANIMAL GROUP** | Reptiles |
| **LENGTH/WEIGHT** | up to 8 ft./120 lb. |
| **DIET** | carnivore: grasshoppers, crickets, amphibians, carrion, fruit flies |
| **LOCATION** | Central and South America |
| **STATUS** | Least Concern |

• March 15th •

# GREAT HAMMERHEAD SHARK

Traveling slowly through the water, the shark is using its well-positioned eyes with 360-degree vision to spot something to eat. It also has electrical sensors in its wide hammer-shaped head to help it find prey, even if the prey is hiding in the sand. This efficient hunter is particularly fond of stingrays, and despite their sharp barbs, can use its head to ram into and pin down a stunned ray on the sea floor to eat it with long, serrated teeth. Hammerheads are not preyed on by other animals.

| | |
|---|---|
| **SCIENTIFIC NAME** | *Sphyrna mokarran* |
| **ANIMAL GROUP** | Fish |
| **LENGTH/WEIGHT** | up to 20 ft./990 lb. |
| **DIET** | carnivore: fish, squid, stingrays, lobsters, eels |
| **LOCATION** | Atlantic, Indian and Pacific Oceans |
| **STATUS** | Critically Endangered |

• March 16th •

# COMMON STINGRAY

This winged fish "flies" along the coastal seabed searching for food with an excellent sense of smell and special sensors around its mouth called ampullae of Lorenzini. It has a poisonous barbed spine up to 14 inches long making up one-third of its whiplike tail. Although it is no real help when it is trying to defend itself against the great hammerhead, the barb will help fend off other sharks and large fish, seals and sea lions.

| | |
|---|---|
| **SCIENTIFIC NAME** | *Dasyatis pastinaca* |
| **ANIMAL GROUP** | Fish |
| **WINGSPAN/WEIGHT** | up to 4.6 ft./70 lb. |
| **DIET** | carnivore: fish, crustaceans, mollusks, worms |
| **LOCATION** | Atlantic Ocean, Mediterranean and Black Seas |
| **STATUS** | Vulnerable |

# CLAWS AND TALONS

Many wild animals have sharp claws or talons that are vital for their existence. They may use them to grasp or kill prey, or they may need them to dig, climb trees or hang from branches.

## • March 17th •
### HARPY EAGLE

At 5 inches, this bird of prey's curved talons are longer than a grizzly bear's claws (see pp.86–87). This heavy and powerful eagle can reach speeds of 50 mph diving onto prey and snatching it up with its talons.

| | |
|---|---|
| SCIENTIFIC NAME | Harpia harpyja |
| ANIMAL GROUP | Birds |
| WINGSPAN/WEIGHT | up to 7 ft./20 lb. |
| DIET | carnivore: opposoms, primates, anteaters |
| LOCATION | northern and central South America |
| STATUS | Vulnerable |

## • March 18th •
### FIDDLER CRAB

The male crab has one enormous claw which it waves to attract females and threaten any rivals. It also uses the claw to fight other males over any burrows it has dug.

| | |
|---|---|
| SCIENTIFIC NAME | Gelasimus tetragonon |
| ANIMAL GROUP | Invertebrates |
| WIDTH | up to 0.8 in. across carapace |
| DIET | omnivore: algae, crabs, larvae, bacteria |
| LOCATION | Asia, South Africa, Australia |
| STATUS | Not Evaluated |

## • March 19th •
### AMERICAN ALLIGATOR

The teeth of this alligator are threatening but it is armed with another weapon: long claws. These claws are also used to excavate gator holes in the mud that fill with water and protect it from both hot and cold weather.

| | |
|---|---|
| SCIENTIFIC NAME | Alligator mississippiensis |
| ANIMAL GROUP | Reptiles |
| LENGTH/WEIGHT | up to 15 ft./990 lb. |
| DIET | carnivore: fish, snails, birds, frogs, mammals |
| LOCATION | southeastern North America |
| STATUS | Least Concern |

• March 20th •
# GIANT ARMADILLO

With the powerful central claws on its front feet, this animal rips open termite mounds and digs for other prey. It has extremely long claws – one-fifth of its body length.

| | |
|---|---|
| SCIENTIFIC NAME | *Priodontes maximus* |
| ANIMAL GROUP | Mammals |
| LENGTH/WEIGHT | up to 5 ft. including tail/130 lb. |
| DIET | insectivore: termites, ants, worms |
| LOCATION | northern and central South America |
| STATUS | Vulnerable |

• March 21st •
# BROWN-THROATED SLOTH

Hanging upside down all day in the trees, sloths rely on large, curved claws to grasp branches and vines. The claws allow them to hang safely and comfortably while they eat and sleep.

| | |
|---|---|
| SCIENTIFIC NAME | *Bradypus variegatus* |
| ANIMAL GROUP | Mammals |
| LENGTH/WEIGHT | up to 2.6 ft. including tail/14 lb. |
| DIET | herbivore: leaves, twigs, fruit |
| LOCATION | central and southern South America |
| STATUS | Least Concern |

• March 22nd •
# LEOPARD

Also called a panther, this large cat has sharp, needlelike claws that are 1 inch long. They are used to fight and to grasp its prey while it kills with its teeth. The leopard takes dead prey up trees to prevent it being stolen by lions or hyenas.

| | |
|---|---|
| SCIENTIFIC NAME | *Panthera pardus* |
| ANIMAL GROUP | Mammals |
| LENGTH/WEIGHT | up to 10 ft. including tail/200 lb. |
| DIET | carnivore: jackals, antelopes, snakes, impala, gazelles, monkeys |
| LOCATION | eastern Africa, Asia |
| STATUS | Vulnerable |

• March 23rd •

## NUDIBRANCH

Also known as the blue sea dragon, this is a hungry sea slug that preys on animals much larger than itself, including the highly venomous Portuguese man o' war (*see p.19*). The nudibranch is not venomous itself, but it swallows and stores the stinging nematocysts of the man o' war in the tips of its extremities. It uses these stinging "fingers" to defend itself when necessary.

| | |
|---|---|
| **SCIENTIFIC NAME** | *Glaucus atlanticus* |
| **ANIMAL GROUP** | Invertebrates |
| **LENGTH/WEIGHT** | up to 1.2 in./1.8 oz. |
| **DIET** | carnivore: Portuguese man o' war, violet snails, other stinging animals |
| **LOCATION** | Atlantic, Pacific and Indian Oceans |
| **STATUS** | Not Evaluated |

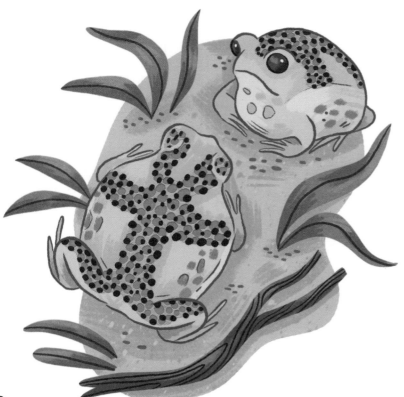

• March 24th •

## CRUCIFIX FROG

Spending most of the day in deep burrows that it digs with its small feet, this frog emerges at night to find food. If there is a drought, it stays dormant (*see pp.174–175*) underground for sometimes years at a time, keeping moist in a protective cocoon. When rain seeps in, it eats the cocoon and surfaces.

| | |
|---|---|
| **SCIENTIFIC NAME** | *Notaden bennettii* |
| **ANIMAL GROUP** | Amphibians |
| **LENGTH** | up to 2.6 in. |
| **DIET** | carnivore: mosquito larvae, insects, tadpoles, ants |
| **LOCATION** | eastern Australia |
| **STATUS** | Least Concern |

• March 25th •

## GREAT SPOTTED WOODPECKER

In spring, this woodland bird can be heard drumming on trees to establish its territory and attract a mate. It has a strong beak and a shock-absorbing skull. This enables it to make holes in trees to raise young, and also to get insects from under tree bark or drink tree sap. It licks the insects up with a long, bristly tongue that can extend up to 1.6 inches beyond its beak.

| | |
|---|---|
| **SCIENTIFIC NAME** | *Dendrocopos major* |
| **ANIMAL GROUP** | Birds |
| **WINGSPAN/WEIGHT** | up to 15 in./3.5 oz. |
| **DIET** | omnivore: beetles, larvae, caterpillars, spiders, seeds, nuts, birds' eggs |
| **LOCATION** | Europe, Asia |
| **STATUS** | Least Concern |

• March 26th •

## AXOLOTL

Only found in the rivers and lakes of central Mexico, this salamander needs to be surrounded by a lot of aquatic plant life where it can find prey to suck into its mouth. Unlike other amphibians, it lives all its life in the water and, keeping its feathery gills, never develops lungs and legs to go on land. It can also regrow lost limbs or a damaged organ, such as its heart or lungs, in only a few weeks.

| | |
|---|---|
| **SCIENTIFIC NAME** | *Ambystoma mexicanum* |
| **ANIMAL GROUP** | Amphibians |
| **LENGTH/WEIGHT** | up to 9 in./9 oz. |
| **DIET** | carnivore: small fish, worms, crustaceans, insects |
| **LOCATION** | North America |
| **STATUS** | Critically Endangered |

• March 27th •

# COMMON SHREW

These little mammals sniff out food on the ground, making high-pitched sounds as they go. They do not have a long life (barely more than a year), but are very busy, as they need to eat 80–90 percent of their body weight to survive. They alternately search for food and rest in their nest all day long. They are too small to hibernate as they cannot store enough fat.

| | |
|---|---|
| **SCIENTIFIC NAME** | *Sorex araneus* |
| **ANIMAL GROUP** | Mammals |
| **LENGTH/WEIGHT** | up to 3.1 in./0.5 oz. |
| **DIET** | carnivore: earthworms, snails, insects, wood lice, plant matter |
| **LOCATION** | Europe |
| **STATUS** | Least Concern |

• March 28th •

# ORNATE BLUET DAMSELFLY

Perched on a reed above a pond, the damselfly holds its transparent wings together above its body and watches its world with its three simple eyes. It spent five years underwater as a nymph before emerging as this beautiful predator of mosquitoes and midges. To catch its prey, it hovers or flies low over shallow, sunny streams and lakes.

| | |
|---|---|
| **SCIENTIFIC NAME** | *Coenagrion ornatum* |
| **ANIMAL GROUP** | Invertebrates |
| **WINGSPAN/LENGTH** | up to 0.9 in./1.2 in. |
| **DIET** | carnivore: insect larvae, mosquitoes, worms, tadpoles |
| **LOCATION** | Europe, Asia |
| **STATUS** | Near Threatened |

• March 29th •

# NORTHERN RED SALAMANDER

By shooting out and pulling back its tongue in milliseconds, this amphibian gives prey no time to escape. In turn, when the salamander needs to defend itself against a predator, it wraps its tail and hind limbs round its head for protection.

| | |
|---|---|
| **SCIENTIFIC NAME** | *Pseudotriton ruber* |
| **ANIMAL GROUP** | Amphibians |
| **LENGTH/WEIGHT** | up to 7 in. including tail/0.7 oz. |
| **DIET** | carnivore: insects, earthworms, spiders, crustaceans, snails |
| **LOCATION** | southeastern North America |
| **STATUS** | Least Concern |

• March 30th •

# FALSE TOMATO FROG

Only found on the island of Madagascar, this rainforest frog is well camouflaged in leaf litter on the forest floor. When threatened it releases a toxin that tastes bad to the predator.

| | |
|---|---|
| **SCIENTIFIC NAME** | *Dyscophus guineti* |
| **ANIMAL GROUP** | Amphibians |
| **LENGTH/WEIGHT** | up to 4 in./8 oz. |
| **DIET** | carnivore: insects, earthworms |
| **LOCATION** | Africa |
| **STATUS** | Least Concern |

• March 31st •

# TEXAS HORNED LIZARD

This short-tailed lizard is heavily armored with large scales, two of which make the "horns" on the back of its head. It can spew blood from the corner of its eyes to confuse a predator.

| | |
|---|---|
| **SCIENTIFIC NAME** | *Phrynosoma cornutum* |
| **ANIMAL GROUP** | Reptiles |
| **LENGTH/WEIGHT** | up to 5 in./3.2 oz. |
| **DIET** | carnivore: mainly harvester ants, also beetles, grasshoppers, spiders |
| **LOCATION** | North America |
| **STATUS** | Least Concern |

# APRIL

• April 1st •

## RAGGIANA BIRD-OF-PARADISE

This magnificent bird is the national bird of the island country of Papua New Guinea, where some local people exchange its feathers for goods. It lives in lowland forest, where it forages for insects in bark crevices. Each year, several males dance together, making loud calls to attract a female. The most impressive dancer wins the competition.

**SCIENTIFIC NAME** | *Paradisaea raggiana*
**ANIMAL GROUP** | Birds
**LENGTH/WEIGHT** | up to 4 ft. including tail feathers/12 oz.
**DIET** | omnivore: figs, insects, frogs, reptiles
**LOCATION** | southwestern Pacific Ocean
**STATUS** | Least Concern

• April 2nd •

# ASIAN GARDEN DORMOUSE

Also known as the large-eared garden dormouse, this nocturnal hunter searches for snails, centipedes and geckos, but likes fruit, nuts and seeds as well. It is usually active all year in forests and gardens, but will go into a state of torpor (*see pp.174–175),* sometimes for several days, if the weather gets really cold.

**SCIENTIFIC NAME** | *Eliomys melanurus*
**ANIMAL GROUP** | Mammals
**LENGTH/WEIGHT** | up to 11 in. including tail/3.5 oz.
**DIET** | omnivore: insects, snails, geckos, fruit
**LOCATION** | western Asia, Africa
**STATUS** | Least Concern

• April 3rd •

# HAIRY FROGFISH

A very strange-looking fish with a name to match, this master of camouflage can completely change its color to match its surroundings. The spines all over its body look like strands of hair and help it hide from predators. It moves along slowly by gulping in water then forcing it out through its gills but can strike at prey very quickly.

**SCIENTIFIC NAME** | *Antennarius striatus*
**ANIMAL GROUP** | Fish
**LENGTH/WEIGHT** | up to 10 in./1 oz.
**DIET** | carnivore: crustaceans, fish
**LOCATION** | Atlantic, Indian and Pacific Ocean
**STATUS** | Least Concern

• April 4th •

## PHILIPPINE EAGLE

Looking for its monkey prey with sharp eyesight, this eagle soars above the tropical forests of islands in the Philippines. Given its length and the large size of its wings, it is considered to be the largest of all the eagles. It eats monkeys, particularly macaques, but will also eat anything else it can grip with its sharp, curved talons. It is very strong and can carry prey that is twice its own weight.

| | |
|---|---|
| **SCIENTIFIC NAME** | *Pithecophaga jefferyi* |
| **ANIMAL GROUP** | Birds |
| **WINGSPAN/WEIGHT** | up to 7 ft./18 lb. |
| **DIET** | carnivore: flying lemurs, giant cloud rats, deer, monkeys |
| **LOCATION** | Asia |
| **STATUS** | Critically Endangered |

• April 5th •

## GLASS-WINGED BUTTERFLY

This delicate butterfly has transparent wings, which makes camouflaging itself very easy because the wings do not shine or shimmer in sunlight. It drinks nectar from particular flowers and absorbs chemicals from them, making it taste unpleasant and deterring predators. The glass-wing is migratory, traveling great distances – up to 12 miles a day at a speed of up to 8 mph.

| | |
|---|---|
| **SCIENTIFIC NAME** | *Greta oto* |
| **ANIMAL GROUP** | Invertebrates |
| **WINGSPAN** | up to 2.4 in. |
| **DIET** | herbivore: nectar, flowers |
| **LOCATION** | Central and South America |
| **STATUS** | Not Evaluated |

• April 6th •

## AYE-AYE

This large-eyed, long-fingered lemur is only found on the island of Madagascar, off the eastern coast of Africa. It is nocturnal, sleeping during the day in nests that it builds from vines in the forks of trees. At night, it forages either on its own or in a small group of three or four over a wide area in the forests and mangrove swamps where it lives. It uses its long middle fingers to tap on trees, listening for echoes from hollow spaces in which it will be able to find grubs to eat.

| | |
|---|---|
| **SCIENTIFIC NAME** | *Daubentonia madagascariensis* |
| **ANIMAL GROUP** | Mammals |
| **LENGTH/WEIGHT** | up to 3 ft. including tail/6 lb. |
| **DIET** | omnivore: fruit, leaves, buds, insects |
| **LOCATION** | Africa |
| **STATUS** | Endangered |

• April 7th •

## COTTON HARLEQUIN BEETLE

These brightly colored beetles are not trying to hide from anything. This female beetle and her young are easily spotted in the many environments they inhabit – from rainforest to coastal dunes. They mainly feed on hibiscus and cotton plants, piercing the stem to eat the sap.

| | |
|---|---|
| **SCIENTIFIC NAME** | *Tectocoris diophthalmus* |
| **ANIMAL GROUP** | Invertebrates |
| **LENGTH** | up to 0.8 in. |
| **DIET** | herbivore: sap of young shoots |
| **LOCATION** | eastern Australia, Pacific islands |
| **STATUS** | Not Evaluated |

• April 8th •
# BARN OWL

Using its extremely sensitive hearing to pinpoint its prey, this barn owl has swooped down in the dark to catch a vole. It is taking it back to feed young in a nest in a barn or other outbuilding, a tree or on a cliff. It will continue to hunt for its young until they fly, at around 55 days after hatching. An adult owls swallows prey whole, regurgitating fur, bones, teeth and feathers as pellets.

| | |
|---|---|
| **SCIENTIFIC NAME** | *Tyto alba* |
| **ANIMAL GROUP** | Birds |
| **WINGSPAN/WEIGHT** | up to 3 ft./13 oz. |
| **DIET** | carnivore: voles, wood mice |
| **LOCATION** | worldwide except Antarctica |
| **STATUS** | Least Concern |

• April 9th •
# JORO SPIDER

Named after a Japanese spider demon, the female Joro spider may be as big as a person's palm, although the male is much smaller and brown in color. She is venomous and, to catch prey, creates a web that is often shaped like a basket and can be up to 10 feet deep. This large spider also uses its silk threads to catch wind currents and glide long distances.

| | |
|---|---|
| **SCIENTIFIC NAME** | *Trichonephila clavata* |
| **ANIMAL GROUP** | Invertebrates |
| **LEG SPAN** | up to 4 in. |
| **DIET** | carnivore: mosquitoes, stink bugs |
| **LOCATION** | Asia |
| **STATUS** | Least Concern |

# DEEP SEA MYSTERIES

The dangerous and unexplored areas of Earth's deep oceans reveal new species day by day. Even in the wildest reaches of our imaginations, we could not dream of some of the extraordinary animals found there.

• April 10th •
## YETI CRAB

This eyeless, hairy crab needs warmth to survive in the chilly 1.2-mile depths where it lives. So it thrives near hydrothermal vents, where cold saltwater is heated by hot magma from underneath the Earth's crust.

| | |
|---|---|
| SCIENTIFIC NAME | *Kiwa hirsuta* |
| ANIMAL GROUP | Invertebrates |
| LENGTH/WEIGHT | up to 6 in./5 lb. |
| DIET | omnivore: bacteria, mussels |
| LOCATION | Pacific Ocean |
| STATUS | Not Evaluated |

• April 11th •
## BLOBFISH

Described by some as the ugliest of all fish, these animals live off the coast of Australia. They are jellylike "blobs" that float and swim up to 4,000 feet deep. Although they lay thousands of eggs, few survive. Those that do may live for 100 years.

| | |
|---|---|
| SCIENTIFIC NAME | *Psychrolutes marcidus* |
| ANIMAL GROUP | Fish |
| LENGTH/WEIGHT | up to 1 ft./9 lb. |
| DIET | omnivore: crustaceans, brittle stars, plants |
| LOCATION | Pacific Ocean |
| STATUS | Not Evaluated |

• April 12th •
## ARMORED SNAIL

This snail is protected by an outer layer of shell that is made out of tough iron sulphide. It lives on deep-ocean ridges near hydrothermal vents at depths of up to 1.8 miles.

| | |
|---|---|
| SCIENTIFIC NAME | *Chrysomallon squamiferum* |
| ANIMAL GROUP | Invertebrates |
| LENGTH | 1.8 in. |
| DIET | omnivore: bacteria |
| LOCATION | Indian Ocean |
| STATUS | Endangered |

• April 13th •
# VAMPIRE SQUID

In the dark depths, this squid shines with lights at the tips of each of its eight arms. If it feels threatened, it turns its cloak inside out to show large spines on the underside.

| | |
|---|---|
| **SCIENTIFIC NAME** | *Vampyroteuthis infernalis* |
| **ANIMAL GROUP** | Invertebrates |
| **LENGTH/WEIGHT** | up to 1 ft./1 lb. |
| **DIET** | omnivore: plant and animal matter |
| **LOCATION** | warm oceans worldwide |
| **STATUS** | Not Evaluated |

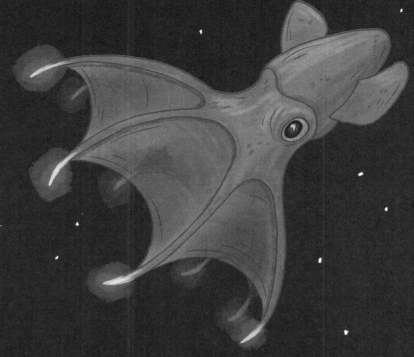

• April 14th •
# HUMPBACK ANGLERFISH

Another sea creature that is bioluminescent – can produce its own light – is this fierce female anglerfish. It waves its light around to attract prey fish, and is able to extend its jaw and stomach to swallow whole prey that is twice its size.

| | |
|---|---|
| **SCIENTIFIC NAME** | *Melanocetus johnsonii* |
| **ANIMAL GROUP** | Fish |
| **LENGTH** | up to 7 in. |
| **DIET** | carnivore: fish, shrimp, squid, turtles |
| **LOCATION** | oceans worldwide |
| **STATUS** | Least Concern |

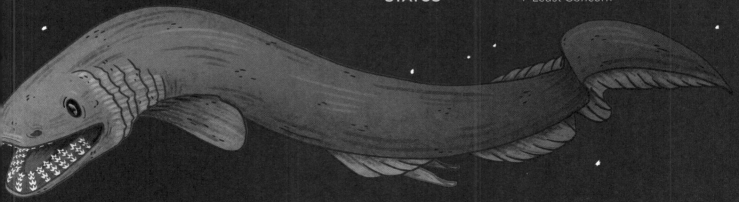

• April 15th •
# FRILLED SHARK

An eellike fish that has changed very little since prehistoric times, this shark is found at depths of 3,200 feet. When it swims, it curls and coils like a snake, pushing its way through the water with its tail.

| | |
|---|---|
| **SCIENTIFIC NAME** | *Chlamydoselachus anguineus* |
| **ANIMAL GROUP** | Fish |
| **LENGTH/WEIGHT** | up to 7 ft./300 lb. |
| **DIET** | carnivore: small fish, squid, other sharks |
| **LOCATION** | Pacific and Atlantic Oceans |
| **STATUS** | Least Concern |

• April 16th •

# SIDEWINDER

The Mojave Desert is home to a pit viper that can move at up to 19 mph. Throwing raised loops of its body to the side to push itself along in an S-shaped curve, this sidewinder is now swinging its fangs forward to stab its prey. The hollow fangs are used to inject a venom that will quickly immobilize the kangaroo rat.

| | |
|---|---|
| **SCIENTIFIC NAME** | *Crotalus cerastes* |
| **ANIMAL GROUP** | Reptiles |
| **LENGTH/WEIGHT** | up to 2.8 ft./11 oz. |
| **DIET** | carnivore: reptiles, rodents, other mammals, birds |
| **LOCATION** | southwestern North America |
| **STATUS** | Least Concern |

• April 17th •

# DESERT KANGAROO RAT

Able to jump up to 9 feet into the air to escape the snake, this rat also needs to move very fast to get away. Safety is a burrow system dug under the sparse vegetation of the hot desert. During the day, the kangaroo rat seals off a den to sleep in, but at night, it needs to feed and so often finds itself in danger.

| | |
|---|---|
| **SCIENTIFIC NAME** | *Dipodomys deserti* |
| **ANIMAL GROUP** | Mammals |
| **LENGTH/WEIGHT** | up to 1 ft. including tail/5 oz. |
| **DIET** | herbivore: seeds, dried plants |
| **LOCATION** | southwestern North America |
| **STATUS** | Least Concern |

## • April 18th •
## LUNA MOTH

This beautiful silk moth is a strong flier in the evenings, and like other moths, is attracted to lights. The false eyes, called ocelli, on its four wings deter predators. It has no mouth because it does not eat, so only lives for about a week. After mating, the female lays 200 eggs at a time in clumps on a leaf before dying.

| | |
|---|---|
| **SCIENTIFIC NAME** | *Actias luna* |
| **ANIMAL GROUP** | Invertebrates |
| **WINGSPAN/WEIGHT** | up to 4.3 in./0.1 oz. |
| **DIET** | herbivore: leaves (when caterpillars) |
| **LOCATION** | North America |
| **STATUS** | Not Evaluated |

## • April 19th •
## PEACOCK MANTIS SHRIMP

A colorful crustacean that lives on the sandy sea floor of coral reefs, the mantis hides away in crevices until prey gets close. Then it is both aggressive and deadly. Very fast, it is able to travel 30 times its own body length in only one second and hit prey, such as a crab, with a force that smashes its shell. And the speed of the mantis's punch is so great that it creates bubbles in the water as hot as the sun, causing a shockwave strong enough to stun or kill.

| | |
|---|---|
| **SCIENTIFIC NAME** | *Odontodactylus scyllarus* |
| **ANIMAL GROUP** | Invertebrates |
| **LENGTH/WEIGHT** | up to 7 in./2.2 oz. |
| **DIET** | carnivore: fish, crabs, worms, shrimps |
| **LOCATION** | Pacific and Indian Oceans |
| **STATUS** | Not Evaluated |

• April 20th •

# STELLER'S JAY

With its perky black crest, this jay is easily recognized flying through coniferous woods. It thrives on pine seeds, likes acorns, and in the summer eats beetles, wasps and wild bees as well. These jays are noisy, making harsh calls and soft whistles. They are very social, forming long-lasting pairs and often gathering in flocks. They also play with or chase each other while flying.

| | |
|---|---|
| SCIENTIFIC NAME | *Cyanocitta stelleri* |
| ANIMAL GROUP | Birds |
| WINGSPAN/WEIGHT | up to 1.4 ft./5 oz. |
| DIET | omnivore: seeds, nuts, insects |
| LOCATION | western North and Central America |
| STATUS | Least Concern |

• April 21st •

# PINE MARTEN

An agile animal and great climber, the pine marten is well adapted to living in forests. It has powerful forelimbs, a long, bushy tail for balance and long, sharp claws to grip. It prefers forests but in some places is found in shrubland or on moors where there is little tree cover. Solitary, it establishes its territory by scent-marking – signaling by scent that intruders are not welcome. It makes a den to sleep in during the day in hollow trees or under the roots of pines. It may travel up to 19 miles a night in search of prey.

| | |
|---|---|
| SCIENTIFIC NAME | *Martes martes* |
| ANIMAL GROUP | Mammals |
| LENGTH/WEIGHT | up to 2.7 ft. including tail/4.9 lb. |
| DIET | omnivore: rodents, squirrels, rabbits, fish, fruit |
| LOCATION | northern Europe, western Asia |
| STATUS | Least Concern |

### • April 22ⁿᵈ •

# HUMPBACK WHALE

Why humpbacks breach – leaping out of the water and falling back with an enormous splash – is a subject much debated by scientists. Some say it is to get rid of itchy parasites, some that it is a form of communication and others that it may just be for fun! These magnificent whales feed on krill and small fish, straining them from the water through baleen plates.

| | |
|---|---|
| **SCIENTIFIC NAME** | *Megaptera novaeangliae* |
| **ANIMAL GROUP** | Mammals |
| **LENGTH/WEIGHT** | up to 60 ft./40 tons |
| **DIET** | carnivore: plankton, small fish, crustaceans |
| **LOCATION** | all oceans |
| **STATUS** | Least Concern |

• April 23rd •

## ALPINE NEWT

These salamanders live in European forests near water, either up in the mountains or down in the valleys. They spend most of the year on land, but every spring return to the water to breed. Females deposit up to 200 eggs, singly or in short chains of 3 to 5 eggs under leaves in shallow ponds or slow-moving streams. The larvae eat tiny crustaceans and aquatic insects, growing up to 2 inches in three months. They then metamorphose into "efts," young versions of the adults, and move to living on land. All salamanders have the amazing ability to regrow parts of their bodies if they are injured.

| | |
|---|---|
| **SCIENTIFIC NAME** | *Ichthyosaura alpestris* |
| **ANIMAL GROUP** | Amphibians |
| **LENGTH/WEIGHT** | up to 4.7 in./0.2 oz. |
| **DIET** | carnivore: amphibian eggs and larvae, insects, worms |
| **LOCATION** | Europe |
| **STATUS** | Least Concern |

• April 24th •

## EMERALD TREE BOA

Deep in the rainforests of the Amazon lives a slender, powerful snake that is an effective ambush hunter. The boa has a long tail that is prehensile, adapted to grasp or hold onto the branches of the trees it inhabits. Its sharp eyes spot any movement, and pits in the scales round its mouth sense the warm blood of any nearby animal. Its teeth point backward so prey cannot escape easily once it is caught.

| | |
|---|---|
| **SCIENTIFIC NAME** | *Corallus caninus* |
| **ANIMAL GROUP** | Reptiles |
| **LENGTH/WEIGHT** | up to 8 ft./4.4 lb. |
| **DIET** | carnivore: rodents, monkeys, bats |
| **LOCATION** | northern South America |
| **STATUS** | Least Concern |

# MIMICRY

Copying a different species is a strategy used by many different animals. It may be the only way to avoid a predator or attract a mate. Changing color or shape may help them blend into the background, or they may actually alter their appearance.

## • April 25th •
### OWL BUTTERFLY

When this butterfly closes its wings, the two large "eyes" on the brown undersides can fool bird predators into thinking it is an owl.

| | |
|---|---|
| SCIENTIFIC NAME | *Caligo memnon* |
| ANIMAL GROUP | Invertebrates |
| WINGSPAN | up to 6 in. |
| DIET | herbivore: rotting fruit, nectar |
| LOCATION | Central and South America |
| STATUS | Not Evaluated |

## • April 26th •
### SEYCHELLES LEAF INSECT

Leaf and stick insects look like the plants they live on and this can protect them from predators. This male leaf insect is shaped like a collection of leaves and is very well camouflaged.

| | |
|---|---|
| SCIENTIFIC NAME | *Pulchriphyllium bioculatum* |
| ANIMAL GROUP | Invertebrates |
| LENGTH | up to 3.7 in. |
| DIET | herbivore: leaves |
| LOCATION | Asia |
| STATUS | Not Evaluated |

## • April 27th •
### SUPERB LYREBIRD

Song and dance are used by this champion mimic to woo a female, and it also copies anything it hears – other birds, fluttering wings, car alarms, camera shutters, even chainsaws!

| | |
|---|---|
| SCIENTIFIC NAME | *Menura novaehollandiae* |
| ANIMAL GROUP | Birds |
| LENGTH/WEIGHT | up to 3.3 ft. including tail/2.4 lb. |
| DIET | omnivore: worms, insects, fungi |
| LOCATION | Australia |
| STATUS | Least Concern |

• April 28th •

## SPHINX HAWK MOTH

The inch-long caterpillar of the sphinx hawk moth has a unique stage in its development. It imitates a deadly pit viper to protect itself against birds and other predators. It twists to show its underside, pulls in its legs and pumps air into itself to expand its body, so it looks just like a snake.

| | |
|---|---|
| SCIENTIFIC NAME | *Hemeroplanes triptolemus* |
| ANIMAL GROUP | Invertebrates |
| WINGSPAN | up to 3.7 in. |
| DIET | herbivore: nectar, fruit |
| LOCATION | North America |
| STATUS | Not Evaluated |

• April 29th •

## EASTERN HOGNOSE

If this mildly venomous snake is confronted by a raccoon or opossum, it flattens its head and neck, hisses and makes moves as if it will strike, behaving like a cobra. If this does not work, it rolls over, sticks out its tongue and plays dead.

| | |
|---|---|
| SCIENTIFIC NAME | *Heterodon platirhinos* |
| ANIMAL GROUP | Reptiles |
| LENGTH/WEIGHT | up to 3.9 ft./12 oz. |
| DIET | carnivore: toads, mammals, birds |
| LOCATION | eastern North America |
| STATUS | Least Concern |

• April 30th •

## MIMIC OCTOPUS

If threatened, this intelligent octopus can make itself look like a fish, pulling itself into a flat shape and jet-propelling itself through the water. It can also change color to yellow and black bands and wave two of its arms out of a burrow as if it were a sea snake. Or it can trail its arms like the poisonous fins of a lionfish.

| | |
|---|---|
| SCIENTIFIC NAME | *Thaumoctopus mimicus* |
| ANIMAL GROUP | Invertebrates |
| LENGTH/WEIGHT | up to 2 ft./20 lb. |
| DIET | carnivore: small crustaceans, fish |
| LOCATION | Indian and western Pacific Oceans |
| STATUS | Least Concern |

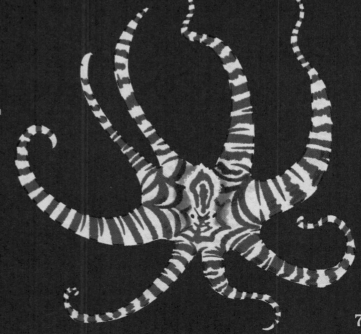

# MAY

• May 1st •

## GREAT WHITE SHARK

With its torpedo shape, powerful tail and bursts of speed, this is the largest predatory fish and a truly terrifying hunter. It is an apex (top) predator that detects prey using special sense organs called ampullae of Lorenzini, before charging at it through the water with lightning speed. Its 300 serrated teeth arranged in up to seven rows are used to rip prey such as this Cape fur seal into bite-size pieces that it swallows whole.

| | |
|---|---|
| SCIENTIFIC NAME | *Carcharodon carcharias* |
| ANIMAL GROUP | Fish |
| LENGTH/WEIGHT | up to 20 ft./2.5 tons |
| DIET | carnivore: seals, sea lions, tuna, squid, other sharks |
| LOCATION | oceans worldwide |
| STATUS | Vulnerable |

• May 2nd •

## CAPE FUR SEAL

Also known as the brown fur seal, this is the largest member of the fur seal family. These eared seals can walk on land but spend weeks at a time hunting for food along the coast of southwest South Africa and southeast Australia.

**SCIENTIFIC NAME** | *Arctocephalus pusillus*
**ANIMAL GROUP** | Mammals
**LENGTH/WEIGHT** | up to 8 ft./795 lb.
**DIET** | carnivore: pilchards, seagulls, penguins, squid
**LOCATION** | Indian and Pacific Oceans
**STATUS** | Least Concern

• May 3rd •

## NILE CROCODILE

The second-largest reptile after the saltwater crocodile (*see p.31*) begins life inside an egg laid in a shallow nest near a river. Although the mother Nile crocodile has jaws that can crush the bones of prey, she is very gentle when carrying her offspring to the water after they hatch. She will guard the hatchlings for up to two years, but the young crocodiles hunt for small fish and invertebrates immediately.

**SCIENTIFIC NAME** | *Crocodylus niloticus*
**ANIMAL GROUP** | Reptiles
**LENGTH/WEIGHT** | up to 20 ft. including tail/1,650 lb.
**DIET** | carnivore: fish, zebra, wildebeest
**LOCATION** | Africa
**STATUS** | Least Concern

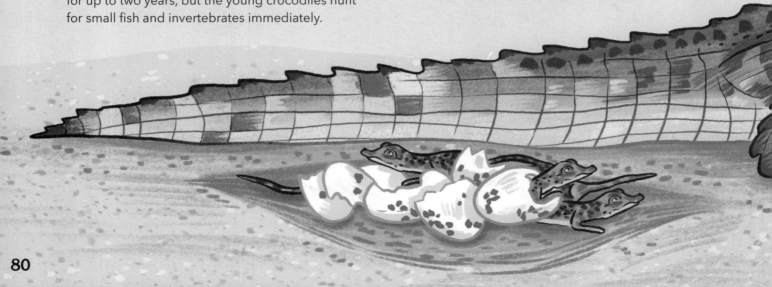

• May 4th •

# BLUE-TAILED BEE-EATER

Calling as they fly on broad, sharply pointed wings over the wetlands and forests where they live, these slender birds take insects on the wing. They carry their catch to a tree and beat it on the perch to break its exoskeleton before eating. This bird is migratory but stays close to water. It nests mostly in river valleys where it tunnels into sand banks.

| | |
|---|---|
| **SCIENTIFIC NAME** | *Merops philippinus* |
| **ANIMAL GROUP** | Birds |
| **WINGSPAN/WEIGHT** | up to 1 ft./1.5 oz. |
| **DIET** | carnivore: bees, hornets, dragonflies |
| **LOCATION** | southeastern Asia |
| **STATUS** | Least Concern |

## • May 5th •
## LAR GIBBON

High up in the trees, this tailless primate brachiates, using only its arms to move up to 50 feet in a single swing from branch to branch in the rainforest. It lives entirely in the upper canopy and never descends to the forest floor. Figs make up half its diet and it uses its hands to scoop water from tree hollows. Lar gibbons live together in small family groups and howl loudly to protect their territory. They spend the night in the forks of "sleeping trees."

| | |
|---|---|
| SCIENTIFIC NAME | *Hylobates lar* |
| ANIMAL GROUP | Mammals |
| LENGTH/WEIGHT | up to 2 ft./17 lb. |
| DIET | omnivore: seeds, figs, leaves, insects |
| LOCATION | Asia |
| STATUS | Endangered |

## • May 6th •
## SACRED SCARAB BEETLE

This dung beetle rolls up balls of dung produced by other animals to feed its young. The ball may be up to 50 times heavier than the beetle itself, and it rolls the dung walking backward with its head near the ground. It was worshipped by the ancient Egyptians both because it represented the god Ra rolling the sun across the sky every day and also as a symbol of rebirth after death.

| | |
|---|---|
| SCIENTIFIC NAME | *Scarabaeus sacer* |
| ANIMAL GROUP | Invertebrates |
| LENGTH/WEIGHT | up to 1.2 in./1.4 oz. |
| DIET | omnivore: dung |
| LOCATION | all continents except Antarctica |
| STATUS | Not Evaluated |

• May 7th •

# YELLOW-BANDED SWEETLIPS

These brightly-colored fish live in shoals in the shallow waters of coral reefs. During the day, they rest in groups under overhanging rocks or large pieces of coral. At night, they emerge to search for food by sifting through mouthfuls of sand for burrowing worms, small bivalves, snails and shrimps.

| | |
|---|---|
| **SCIENTIFIC NAME** | *Plectorhinchus lineatus* |
| **ANIMAL GROUP** | Fish |
| **LENGTH** | up to 20 in. |
| **DIET** | carnivore: invertebrates |
| **LOCATION** | western Pacific Ocean, eastern Indian Ocean |
| **STATUS** | Not Evaluated |

• May 8th •

# MONTANE HOURGLASS TREE FROG

Only found in the central hills of the island of Sri Lanka, this amphibian lives in the canopy of the cloud forests where there is high rainfall, or in the reeds and grasses at the edge of water. Some, like this one, have an hourglass pattern on their backs.

| | |
|---|---|
| **SCIENTIFIC NAME** | *Taruga eques* |
| **ANIMAL GROUP** | Amphibians |
| **LENGTH** | up to 2.8 in. |
| **DIET** | carnivore: insects |
| **LOCATION** | Asia |
| **STATUS** | Endangered |

• May 9th •

# GIANT FOREST ANT

To defend themselves against predators or rivals, these ants bite and use hairs to "paint" formic acid on the attacker. Colonies may contain up to 7,000 ants occupying several nests, with only a single queen in one of the nests. The ants forage for insect honeydew through the rainforest canopy.

| | |
|---|---|
| **SCIENTIFIC NAME** | *Dinomyrmex gigas* |
| **ANIMAL GROUP** | Invertebrates |
| **LENGTH** | up to 1.4 in. |
| **DIET** | omnivores: bird droppings, honeydew, termites |
| **LOCATION** | southeastern Asia |
| **STATUS** | Not Evaluated |

# MIGRATION

Moving from one place to another with the change of seasons each year is called migration. Animals migrate for a variety of reasons – they may be in search of food and water, a place to breed or shelter from the winter cold.

### • May 10th •
## CANADA GOOSE

Wintering in the south, these geese fly north in a V-formation in the spring to breed and nest. They may travel more than 1,000 miles in a single day at speeds of up to 70 mph. They mate for life and nest where their parents nested.

| | |
|---|---|
| SCIENTIFIC NAME | *Branta canadensis* |
| ANIMAL GROUP | Birds |
| WINGSPAN/WEIGHT | up to 6 ft./11 lb. |
| DIET | omnivore: grass, shoots, seeds, berries, insects |
| LOCATION | North America, northern Europe |
| STATUS | Least Concern |

### • May 11th •
## MONARCH BUTTERFLY

In North America, these butterflies travel over 3,000 miles from Mexico to Canada to breed. The journey there and back takes so long that no butterfly makes the round trip.

| | |
|---|---|
| SCIENTIFIC NAME | *Danaus plexippus* |
| ANIMAL GROUP | Invertebrates |
| WINGSPAN/WEIGHT | up to 4 in./0.02 oz. |
| DIET | herbivore: nectar from flowers |
| LOCATION | North America, Central America |
| STATUS | Least Concern |

### • May 12th •
## RUBY-THROATED HUMMINGBIRD

This tiny bird builds a nest the size of a thimble on top of a branch in Canada or eastern North America in the summer. It then beats its wings more than 50 times a second to carry it all the way to Central America, Mexico and Florida, where it spends the winter.

| | |
|---|---|
| SCIENTIFIC NAME | *Archilochus colubris* |
| ANIMAL GROUP | Birds |
| WINGSPAN/WEIGHT | up to 4.3 in./0.2 oz. |
| DIET | omnivore: nectar, insects |
| LOCATION | Central America, North America |
| STATUS | Least Concern |

• May 13th •

# SOCKEYE SALMON

Spending most of their life at sea, these salmon return yearly to freshwater to spawn, traveling upstream to the river in which they hatched. The females dig nests with their tails and deposit their eggs. They cover the eggs with gravel after the males fertilize them. The fry stay up to three years in freshwater before migrating to the sea.

| | |
|---|---|
| SCIENTIFIC NAME | *Oncorhynchus nerka* |
| ANIMAL GROUP | Fish |
| LENGTH/WEIGHT | up to 2.8 ft./15 lb. |
| DIET | carnivore: zooplankton, fish, insects |
| LOCATION | northern Pacific Ocean and rivers |
| STATUS | Least Concern |

• May 14th •

# LEATHERBACK TURTLE

The leatherback is the largest turtle in the world. To reach the tropical beaches where they nest each year, some of these turtles swim more than 10,000 miles. They dig into the sand, lay their eggs, cover them up and return to the sea, leaving the young to hatch, dig themselves out and dash to the ocean.

| | |
|---|---|
| SCIENTIFIC NAME | *Dermochelys coriacea* |
| ANIMAL GROUP | Reptiles |
| LENGTH/WEIGHT | up to 7 ft./1,500 lb. |
| DIET | omnivore: jellyfish, seaweeds, fish |
| LOCATION | oceans worldwide except Arctic and Southern |
| STATUS | Vulnerable |

• May 15th •

# GRAY WHALE

Making one of the longest migrations of any mammal, the gray whale travels up to 14,000 miles in a round trip. It moves from its summer feeding grounds in the Arctic to winter in calving lagoons near the Equator. The southward journey takes two to three months.

| | |
|---|---|
| SCIENTIFIC NAME | *Eschrichtius robustus* |
| ANIMAL GROUP | Mammals |
| LENGTH/WEIGHT | up to 50 ft./40 tons |
| DIET | carnivore: crustaceans, worms, herring eggs |
| LOCATION | Pacific Ocean |
| STATUS | Least Concern |

• May 16th •

# GRIZZLY BEAR

On their way up the river to spawn, sockeye salmon (see p.85) have to overcome many obstacles, not least the fact they are swimming against the flow of the water. Perhaps the biggest obstacle of all is what happens when these salmon try to leap up waterfalls on the west coast of southern Alaska. Grizzly bears have only to stand at the top with their mouths open to feast on up to 30 big fish a day in June and July. They usually carry the fish to calmer water or the riverbank to eat them.

| | |
|---|---|
| **SCIENTIFIC NAME** | Ursus arctos horribilis |
| **ANIMAL GROUP** | Mammals |
| **LENGTH/WEIGHT** | up to 10 ft. including tail/1,700 lb. |
| **DIET** | omnivore: Arctic fox, salmon, grass, berries |
| **LOCATION** | North America, Asia, Europe |
| **STATUS** | Least Concern |

• May 17th •

# RAINBOW LORIKEET

Unlike most birds, the male and female rainbow lorikeet are a matched pair – both unmistakably and brightly multicolored in the same way. They pair for life and are often seen in loud and fast-moving flocks that nest together at night. The end of their tongues is brush-shaped and covered with little bumps or papillae, so it is easy for them to collect nectar and pollen from flowers.

| | |
|---|---|
| **SCIENTIFIC NAME** | *Trichoglossus moluccanus* |
| **ANIMAL GROUP** | Birds |
| **WINGSPAN/WEIGHT** | up to 18 in./6 oz. |
| **DIET** | omnivore: nectar, pollen, seeds, insects, fruit |
| **LOCATION** | northern and eastern Australia |
| **STATUS** | Least Concern |

• May 18th •

# CHIMPANZEE

One of the great apes, this is one of our closest relatives – we share 98 percent of our DNA with it. Chimpanzees are highly intelligent and regularly employ tools. They use sticks to get ants or termites out of their nests, and smash nuts open with stones. Scientists have also observed sharpening sticks with their teeth to use them as spears to pry bush babies from tree cavities. These apes mainly live in rainforest areas as they need access to water and the fruit they like to eat.

| | |
|---|---|
| **SCIENTIFIC NAME** | *Pan troglodytes* |
| **ANIMAL GROUP** | Mammals |
| **LENGTH/WEIGHT** | up to 6 ft./155 lb. |
| **DIET** | omnivore: fruit, plants, insects, birds, mammals |
| **LOCATION** | eastern and central Africa |
| **STATUS** | Endangered |

• May 19th •

## GROUND PANGOLIN

Covered in protective overlapping scales, this large mammal rolls itself into a ball if threatened, wrapping its tail tightly round its body. It leads a solitary life and is active at night. The large claws on its front feet are used to break into ant hills and termite mounds, where its long, sticky tongue laps up to 70 million insects in a year.

| | |
|---|---|
| SCIENTIFIC NAME | *Smutsia temminckii* |
| ANIMAL GROUP | Mammals |
| LENGTH/WEIGHT | up to 4.6 ft. including tail/40 lb. |
| DIET | carnivore: ants, termites |
| LOCATION | southern and eastern Africa |
| STATUS | Vulnerable |

• May 20th •

## KAISER'S MOUNTAIN NEWT

This slender amphibian is only found in the Zagros Mountains of Iran above 3,300 feet. It lives in woodland near the springs, waterfalls and cold mountain streams where it goes to breed. Its large eyes help it forage for food at night. If it becomes too hot in summer, the newt goes underground where it is cooler and estivates *(see pp. 174–175)*.

| | |
|---|---|
| SCIENTIFIC NAME | *Neurergus kaiseri* |
| ANIMAL GROUP | Reptiles |
| LENGTH | up to 6 in. |
| DIET | carnivore: small invertebrates |
| LOCATION | Asia |
| STATUS | Vulnerable |

# DESERT SURVIVAL

To live in a hot, dry desert, an animal must deal with a lack of water and extremes of temperature – it may be too hot during the day and too cold at night. Many have found unique ways to survive in these conditions.

• May 21st •

## NAMIB DARKLING BEETLE

This little beetle of the Namib Desert uses its body to collect water from fog. It leans into the early morning wind, letting droplets of fog drip down its wing cases into its mouth.

| | |
|---|---|
| SCIENTIFIC NAME | *Stenocara gracilipes* |
| ANIMAL GROUP | Invertebrates |
| LENGTH | up to 1 in. |
| DIET | herbivore: plant debris |
| LOCATION | southern Africa |
| STATUS | Not Evaluated |

• May 22nd •

## DEVIL SCORPION

In the Sonoran Desert, temperatures can reach 118°F. To avoid the extreme heat of the day, the scorpion shelters under rocks and only emerges at night when it is cooler to hunt for food.

| | |
|---|---|
| SCIENTIFIC NAME | *Paravaejovis spinigerus* |
| ANIMAL GROUP | Invertebrates |
| LENGTH/WEIGHT | up to 2.8 in. including tail/0.3 oz. |
| DIET | carnivore: crickets, other scorpions |
| LOCATION | North America |
| STATUS | Not Evaluated |

• May 23rd •

## WILD BACTRIAN CAMEL

The Gobi and Gashan Deserts of China and Mongolia are harsh places, but this camel is well adapted. Its two humps store energy-rich fat to use when there is little food available.

| | |
|---|---|
| SCIENTIFIC NAME | *Camelus ferus* |
| ANIMAL GROUP | Mammals |
| LENGTH/WEIGHT | up to 11 ft./1,500 lb. |
| DIET | herbivore: grasses, leaves, grains |
| LOCATION | Asia |
| STATUS | Critically Endangered |

## • May 24th •
## FENNEC FOX

This desert fox of the Sahara Desert has unusually large ears which help it lose body heat and keep cool. The ears are also very useful to pick up sounds of prey in the sand.

| | |
|---|---|
| **SCIENTIFIC NAME** | *Vulpes zerda* |
| **ANIMAL GROUP** | Mammals |
| **LENGTH/WEIGHT** | up to 2 ft. including tail/3 lb. |
| **DIET** | omnivore: grasshoppers, locusts, rodents, fruit, leaves |
| **LOCATION** | northern Africa |
| **STATUS** | Least Concern |

## • May 25th •
## THORNY DEVIL

Active during the day, this reptile has only to stand still or rub against dew-covered plants to collect water. The scales on its body have ridges that are shaped to direct water to its mouth.

| | |
|---|---|
| **SCIENTIFIC NAME** | *Moloch horridus* |
| **ANIMAL GROUP** | Reptiles |
| **LENGTH/WEIGHT** | up to 8 in. including tail/3.3 oz. |
| **DIET** | carnivore: black ants |
| **LOCATION** | Australia |
| **STATUS** | Least Concern |

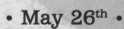

## • May 26th •
## ARABIAN ORYX

Oryx dig into the sand and lie down to cool their bodies by transferring heat into the sand. They are also protected in this way from the dry winds of the desert.

| | |
|---|---|
| **SCIENTIFIC NAME** | *Oryx leucoryx* |
| **ANIMAL GROUP** | Mammals |
| **LENGTH/WEIGHT** | up to 7.5 ft. including tail/150 lb. |
| **DIET** | herbivore: grass, roots, tubers |
| **LOCATION** | Asia |
| **STATUS** | Vulnerable |

• May 27th •

## ANNA'S EIGHTY-EIGHT BUTTERFLY

Taking its name from the apparent number 88 on the underside of its wings, this beautiful butterfly lives in wet tropical forests. Its caterpillars feed on the leaves of various tropical plants, while the adults eat rotting fruit and dung. It is found all the way from southern Texas to the Brazilian Amazon, and is considered by many Central American cultures as a symbol of good luck.

| | |
|---|---|
| **SCIENTIFIC NAME** | *Diaethria anna* |
| **ANIMAL GROUP** | Invertebrates |
| **WINGSPAN** | up to 1.6 in. |
| **DIET** | herbivore: rotting fruit, dung |
| **LOCATION** | North, Central and South America |
| **STATUS** | Not Evaluated |

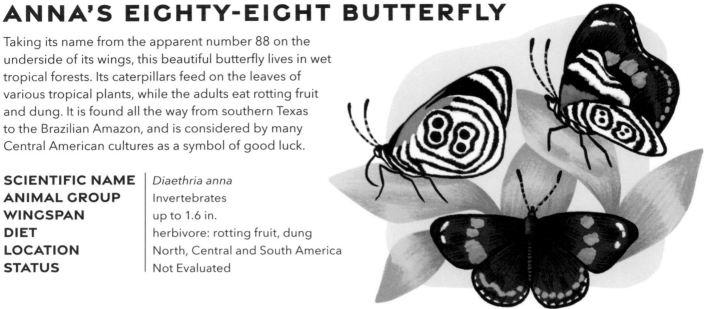

• May 28th •

## PEACOCK WORM

This long, slender worm lives all its life in a flexible tube of mud planted in sand. To feed, its head pokes out from the tube covered with a crown of feathery tentacles to catch particles from the saltwater flowing past and move them to the mouth. The tentacles are then withdrawn.

| | |
|---|---|
| **SCIENTIFIC NAME** | *Sabella pavonina* |
| **ANIMAL GROUP** | Invertebrates |
| **LENGTH/WIDTH** | up to 1 ft./0.2 oz. |
| **DIET** | omnivore: particles from saltwater |
| **LOCATION** | Atlantic Ocean, Mediterranean Sea |
| **STATUS** | Not Evaluated |

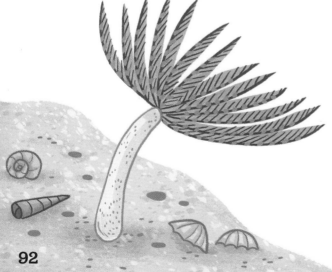

• May 29th •

## REEF STONEFISH

The most venomous fish known, this ray-finned master of disguise injects venom with its dorsal fin spines. It is so well camouflaged it looks just like a rock or a piece of coral and is very hard to spot.

| | |
|---|---|
| **SCIENTIFIC NAME** | *Synanceia verrucosa* |
| **ANIMAL GROUP** | Fishes |
| **LENGTH/WEIGHT** | up to 16 in./5 oz. |
| **DIET** | carnivore: small fish, crustaceans |
| **LOCATION** | western Pacific Ocean, Indian Ocean |
| **STATUS** | Least Concern |

• May 30th •

## MEERKAT

A group of these small mongooses is called a "mob" or a "gang" and is usually made up of meerkats that are related to one another. They live in the Kalahari Desert, where they dig multi-entrance burrows with their long, curved claws. They forage for up to eight hours a day with one acting as a sentinel, and they sleep in the burrow all together.

| | |
|---|---|
| **SCIENTIFIC NAME** | *Suricata suricatta* |
| **ANIMAL GROUP** | Mammals |
| **LENGTH/WEIGHT** | up to 2 ft. including tail/1.6 lb. |
| **DIET** | carnivore: scorpions, insects |
| **LOCATION** | southern Africa |
| **STATUS** | Least Concern |

• May 31st •

## NORTHERN ROCKHOPPER PENGUIN

One of the smallest penguins, this has a wild yellow-and-black crest and red eyes. It can be very loud when fighting for nesting areas and mates, making braying and barking sounds while swinging its head and beating its flippers. It is called "rockhopper" because, although it does slide on its belly like other penguins, it also hops over obstacles on the steep, rocky slopes where it breeds.

| | |
|---|---|
| **SCIENTIFIC NAME** | *Eudyptes moseleyi* |
| **ANIMAL GROUP** | Birds |
| **HEIGHT/WEIGHT** | up to 18 in./10 lb. |
| **DIET** | carnivore: fish, octopus, mollusks, crustaceans |
| **LOCATION** | southern Indian and Atlantic Oceans |
| **STATUS** | Endangered |

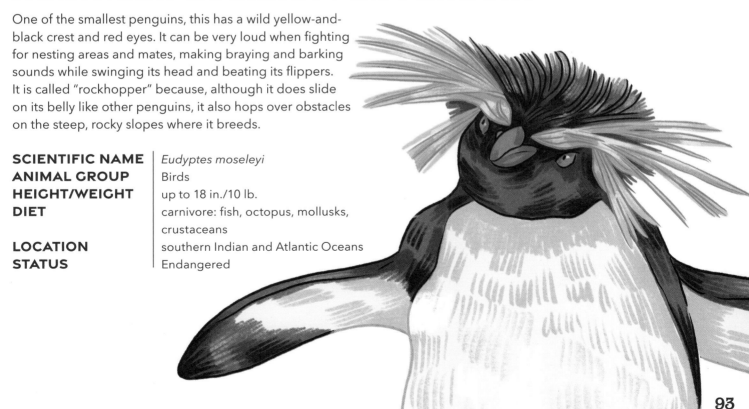

# JUNE

• June 1st •

## GIANT PANDA

To reach the bamboo it depends on to survive, a panda has to be good at climbing. It is a big eater, needing to munch up to 85 pounds of bamboo a day, so it spends up to 16 hours foraging for bamboo shoots, leaves and stems. Pandas are solitary animals, only coming together to mate or raise their young. They live in the Minshan and Qinling Mountains of central China where their thick fur protects them from the misty or snowy conditions in the area.

| | |
|---|---|
| **SCIENTIFIC NAME** | *Ailuropoda melanoleuca* |
| **ANIMAL GROUP** | Mammals |
| **LENGTH/WEIGHT** | up to 6 ft./350 lb. |
| **DIET** | omnivore: mainly bamboo; also fish, small rodents, carrion |
| **LOCATION** | Asia |
| **STATUS** | Vulnerable |

• June 2nd •

# BLUE JAY

Its noisy calls carry over long distances as it flies along forest edges looking for acorns to pluck with its beak. It holds the nut in its claws to peck it open, or stores it in a throat pouch to cache for later. One of its calls mimics (see p.76–77) the scream of the red-shouldered hawk, as a warning or to scare off other birds.

| | |
|---|---|
| **SCIENTIFIC NAME** | *Cyanocitta cristata* |
| **ANIMAL GROUP** | Birds |
| **WINGSPAN/WEIGHT** | up to 1.4 ft./3.5 oz. |
| **DIET** | omnivore: seeds, nuts, berries, ants |
| **LOCATION** | North America |
| **STATUS** | Least Concern |

• June 3rd •

# AARDVARK

A burrowing, nocturnal mammal, the aardvark is the only species in its family. It lives on the savanna south of the Sahara Desert, where it can easily dig into the soil. It gets nearly all its food from underground – ants, termites and other insects – lapping them up with its foot-long tongue. Its name is Afrikaans for "earth pig," and its snout, which it uses to sniff out insects, looks like that of a pig.

| | |
|---|---|
| **SCIENTIFIC NAME** | *Orycteropus afer* |
| **ANIMAL GROUP** | Mammals |
| **LENGTH/WEIGHT** | up to 7 ft. including tail/180 lb. |
| **DIET** | carnivore: insects |
| **LOCATION** | Africa |
| **STATUS** | Least Concern |

• June 4th •

# DAISY PARROTFISH

With its powerful parrotlike beak, this fish tears algae from dead coral to eat. At night, it makes a "sleeping bag" out of its own mucus to cocoon and protect itself from parasites. It is also unusual because it is a hermaphrodite – able to transform from female to male. Females do this if there is no dominant male in a group.

| | |
|---|---|
| **SCIENTIFIC NAME** | *Chlorurus sordidus* |
| **ANIMAL GROUP** | Fish |
| **LENGTH** | up to 1.3 ft. |
| **DIET** | herbivore: algae |
| **LOCATION** | Indian and Pacific Oceans |
| **STATUS** | Least Concern |

• June 5th •

# CAPYBARA

Diving into water to escape jaguars, capybaras are excellent swimmers with webbed feet. They can hold their breath underwater for up to five minutes at a time. These social animals, the largest rodents in the world, also live on land, grazing on grass with long, sharp teeth. If threatened, they can speedily run away at up to 22 mph.

| | |
|---|---|
| **SCIENTIFIC NAME** | *Hydrochoerus hydrochaeris* |
| **ANIMAL GROUP** | Mammals |
| **LENGTH/WEIGHT** | up to 4.3 ft./175 lb. |
| **DIET** | herbivore: plants, grasses, fruit |
| **LOCATION** | South America, east of the Andes |
| **STATUS** | Least Concern |

# LIFE IN FRESHWATER

Only 3 percent of the water on Earth is freshwater, and two-thirds of that is locked up in glaciers and ice at the North and South Poles. However, the rest, in rivers, pools, ponds and lakes, is home to some fascinating animals.

• June 6th •

## GRAY HERON

This heron usually lays its eggs in a large, messy nest of twigs at the top of a tree close to a river. Colonies or "heronries" may number more than 100 nests. The parent herons stand still or walk very slowly through the water nearby to ambush prey. They then regurgitate the fish or frogs to feed their young.

| | |
|---|---|
| **SCIENTIFIC NAME** | *Ardea cinerea* |
| **ANIMAL GROUP** | Birds |
| **WINGSPAN/WEIGHT** | up to 6 ft./4.4 lb. |
| **DIET** | carnivore: fish, amphibians, ducklings |
| **LOCATION** | Europe, Asia, Africa |
| **STATUS** | Least Concern |

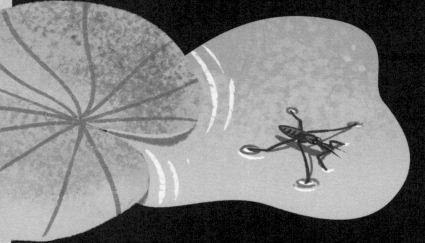

• June 7th •

## COMMON PONDSKATER

A slender insect with six long legs, this "water strider" uses the water's surface tension to skate across streams and lakes. Its body is covered with tiny hairs that repel water.

| | |
|---|---|
| **SCIENTIFIC NAME** | *Gerris lacustris* |
| **ANIMAL GROUP** | Invertebrates |
| **LENGTH** | up to 0.6 in. |
| **DIET** | carnivore: other insects |
| **LOCATION** | Europe, Asia, northern Africa |
| **STATUS** | Not Evaluated |

• June 8th •

## EURASIAN OTTER

With eyes high on its head so it can see what is happening even when the rest of its body is underwater, this large, powerful and playful mammal is an excellent swimmer.

| | |
|---|---|
| **SCIENTIFIC NAME** | *Lutra lutra* |
| **ANIMAL GROUP** | Mammals |
| **LENGTH/WEIGHT** | up to 4.3 ft. including tail/24 lb. |
| **DIET** | carnivore: mainly fish; frogs, birds |
| **LOCATION** | Europe, Asia, northern Africa |
| **STATUS** | Near Threatened |

• June 9th •
# COMMON KINGFISHER

An expert fisher, this bird dives from a perch into slow-moving or still water for small fish, beetles and dragonflies. It captures prey up to 10 inches below the surface, eating 60 percent of its body weight every day. It nests in tunnels in the riverbank.

| | |
|---|---|
| SCIENTIFIC NAME | *Alcedo atthis* |
| ANIMAL GROUP | Birds |
| WINGSPAN/WEIGHT | up to 10 in./1.6 oz. |
| DIET | carnivore: fish, insects |
| LOCATION | Europe, Asia, northern Africa |
| STATUS | Least Concern |

• June 10th •
# BELUGA STURGEON

Female beluga sturgeons are prized for their eggs, which are eaten as caviar. Most of these fish live in the salty Caspian Sea, where their long snouts with their sensitive "whiskers" help them hunt for the fish they eat. The beluga is also found in freshwater, where it is among the largest fish.

| | |
|---|---|
| SCIENTIFIC NAME | *Huso huso* |
| ANIMAL GROUP | Fish |
| LENGTH/WEIGHT | up to 24 ft./1.7 tons |
| DIET | carnivore: fish |
| LOCATION | eastern Europe, western Asia |
| STATUS | Critically Endangered |

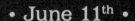

• June 11th •
# AMAZON RIVER DOLPHIN

Also known as the boto, this dolphin is found along most of the Amazon and Orinoco rivers in South America. It is not born pink but changes slowly from gray to pink as it grows. It spends a lot of its time underwater and uses echolocation to navigate away from obstacles and toward prey in the dark waters. Agile, it often swims upside down or leaps out of the water to see what is happening.

| | |
|---|---|
| SCIENTIFIC NAME | *Inia geoffrensis* |
| ANIMAL GROUP | Mammals |
| LENGTH/WEIGHT | up to 9 ft./180 lb. |
| DIET | carnivore: fish, shellfish, crabs, turtles |
| LOCATION | South America |
| STATUS | Endangered |

• June 12th •

# MUTE SWAN

With its long S-shaped neck, this large white waterbird is a familiar sight on rivers and lakes. These swans mate for life, and the males are territorial, attacking rivals and anything they think is a threat to their nests. When the cygnets hatch, they are gray with black beaks, but gradually change color over the following months. The mother is carrying them on her back where they will keep warm and conserve energy. They learn how to swim and feed by watching her.

| | |
|---|---|
| **SCIENTIFIC NAME** | *Cygnus olor* |
| **ANIMAL GROUP** | Birds |
| **WINGSPAN/WEIGHT** | up to 8 ft./26 lb. |
| **DIET** | omnivore: water plants, insects, snails, grass |
| **LOCATION** | Europe, Asia |
| **STATUS** | Least Concern |

• June 13th •
# TOKAY GECKO

One of the largest of the geckos, this reptile can climb up any surface. It is a solitary night hunter with a strong bite. If threatened, it loses its tail, which then regrows.

| | |
|---|---|
| **SCIENTIFIC NAME** | *Gekko gecko* |
| **ANIMAL GROUP** | Reptiles |
| **LENGTH/WEIGHT** | up to 16 in./14 oz. |
| **DIET** | carnivore: insects, spiders, mice, snakes |
| **LOCATION** | southeastern Asia |
| **STATUS** | Least Concern |

• June 14th •
# MANDRILL

It is hard to spot a mandrill as they are very shy. They live in the rainforest, where they walk long distances in groups or "hordes" through the dense vegetation foraging for food – their colors help them to follow one another. At night, their strong arms help them climb trees to sleep.

| | |
|---|---|
| **SCIENTIFIC NAME** | *Mandrillus sphinx* |
| **ANIMAL GROUP** | Mammals |
| **LENGTH/WEIGHT** | up to 3.3 ft. including tail/75 lb. |
| **DIET** | omnivore: seeds, nuts, fruit, small reptiles |
| **LOCATION** | western Africa |
| **STATUS** | Vulnerable |

• June 15th •
# BANDED DEMOISELLE

These insects live in the damp reeds and plants of slow-flowing streams and rivers. The adults have long bodies and hornlike antennae. After mating, the female lays her eggs on plant stems just under the surface. The nymphs live underwater for two years before crawling up aquatic plants, shedding their skin and emerging as adults.

| | |
|---|---|
| **SCIENTIFIC NAME** | *Calopteryx splendens* |
| **ANIMAL GROUP** | Invertebrates |
| **WINGSPAN** | up to 1.8 in. |
| **DIET** | carnivore: other insects |
| **LOCATION** | Europe, Asia |
| **STATUS** | Least Concern |

• June 16th •

# BALD EAGLE

This sea eagle has left its aerie high in a tree to find food to feed its growing nestlings. The bald eagle has been the national emblem of the USA since 1782. It is not bald, but was probably given that name because of its white head. It hunts by flying very low over the sea or land, or watching from a perch and then swooping on its prey. It steals the catches of other birds, particularly that of the osprey, and is also a scavenger, feeding on carrion.

| | |
|---|---|
| **SCIENTIFIC NAME** | *Haliaeetus leucocephalus* |
| **ANIMAL GROUP** | Birds |
| **WINGSPAN/WEIGHT** | up to 8 ft./14 lb. |
| **DIET** | carnivore: fish, birds, mammals |
| **LOCATION** | North America |
| **STATUS** | Least Concern |

# HOME BUILDING

Most animals make homes of different shapes and sizes to suit their needs. But some are truly amazing architects and builders. They employ unusual skills to design and build with a precision that is astonishing.

• June 17th •

## BAYA WEAVER BIRD

These small birds weave the most elaborate nests out of wild grasses. Often built in thorn trees, the entrance tube can be as long as 3 feet to protect the chicks from predators such as snakes and lizards.

| | |
|---|---|
| SCIENTIFIC NAME | *Ploceus philippinus* |
| ANIMAL GROUP | Birds |
| LENGTH/WEIGHT | up to 10 in./1 oz. |
| DIET | omnivore: seeds, grains, wild grasses, insects |
| LOCATION | south and southeastern Asia |
| STATUS | Least Concern |

• June 18th •

## AUSTRALIAN HORNET

This is a "potter" or "mason" wasp and the females are great home-builders. They make mud for their nests themselves, finding water, collecting dirt and mixing the two in their mouths. Then they carry the mud back to construct a mud nest, with a tunnel leading to an interior cell, before perfectly smoothing the outside.

| | |
|---|---|
| SCIENTIFIC NAME | *Abispa ephippium* |
| ANIMAL GROUP | Invertebrates |
| LENGTH | up to 1.2 in. |
| DIET | omnivore: nectar, spiders, caterpillars |
| LOCATION | Australia |
| STATUS | Not Evaluated |

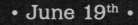

• June 19th •

## NORTH AMERICAN BEAVER

Beavers build great dams, reduce erosion and provide habitats for dozens of other species. They also build domelike wooden lodges that can only be entered from underwater, providing protection for themselves and giving them somewhere to raise their young.

| | |
|---|---|
| SCIENTIFIC NAME | *Castor canadensis* |
| ANIMAL GROUP | Mammals |
| LENGTH/WEIGHT | up to 3.9 ft. including tail/70 lb. |
| DIET | herbivore: leaves, stems, aquatic plants |
| LOCATION | North America |
| STATUS | Least Concern |

• June 20th •

# LEAF-CURLING SPIDER

This fascinating little spider is a member of the orb-weaver family. It protects itself from predators, such as birds or parasitic wasps, by curling a leaf and sewing it together with its silk in the centre of its web.

| | |
|---|---|
| SCIENTIFIC NAME | *Phonognatha graeffei* |
| ANIMAL GROUP | Invertebrates |
| LEGSPAN | up to 1.2 in. |
| DIET | carnivore: insects |
| LOCATION | Australia |
| STATUS | Not Evaluated |

• June 21st •

# VOGELKOP BOWERBIRD

The male bowerbird woos the female by creating an elaborate bower propped up by large sticks. It carefully makes heaps of colorful flowers, fruit, feathers and leaves to attract her.

| | |
|---|---|
| SCIENTIFIC NAME | *Amblyornis inornata* |
| ANIMAL GROUP | Birds |
| LENGTH/WEIGHT | up to 10 in./5 oz. |
| DIET | omnivore: fruit, insects |
| LOCATION | southeastern Asia |
| STATUS | Least Concern |

• June 22nd •

# MEXICAN PRAIRIE DOG

These mammals live underground in large colonies, or towns, of up to 100 animals. They dig burrow systems that may cover 155 square miles. They have chambers for sleeping, storing food and raising young, all connected by a network of tunnels.

| | |
|---|---|
| SCIENTIFIC NAME | *Cynomys mexicanus* |
| ANIMAL GROUP | Mammals |
| LENGTH/WEIGHT | up to 18 in. including tail/4.4 lb. |
| DIET | herbivore: leaves, roots, tubers |
| LOCATION | North America |
| STATUS | Endangered |

• June 23rd •

# COCONUT CRAB

On coral atolls and small islands in the Indian and Pacific Oceans lives a burrowing hermit crab with immensely powerful claws. It is able to exert the strongest crushing force of any crustacean. It climbs trees to eat coconuts, which it opens by cutting holes with its pincers, before breaking the hard shell into pieces with its claws and eating the contents. Although these crabs begin life as larvae in the sea, as adults they cannot swim and will drown in water.

**SCIENTIFIC NAME** | *Birgus latro*
**ANIMAL GROUP** | Invertebrates
**LEGSPAN/WEIGHT** | up to 3.3 ft./9 lb.
**DIET** | omnivore: figs, rats, crustaceans, coconuts, tortoise eggs
**LOCATION** | Indian and western Pacific Oceans
**STATUS** | Vulnerable

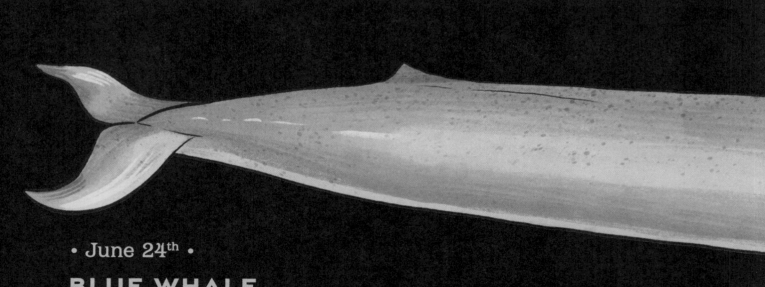

• June 24th •

# BLUE WHALE

This long, streamlined whale is the largest living animal and the largest creature ever to have lived on Earth. It weighs the equivalent of 33 African elephants (see p.118) and its song can be louder than a jet engine. It spends its summers feeding in polar waters, then migrates (see pp.84–85) to winter

**SCIENTIFIC NAME** | *Balaenoptera musculus*
**ANIMAL GROUP** | Mammals
**LENGTH/WEIGHT** | up to 110 ft./210 tons
**DIET** | carnivore: krill
**LOCATION** | oceans worldwide except the Arctic

• June 25th •

# BLUE-FOOTED BOOBY

Boobies have a very wobbly waddle on land, which caused the Spaniards that first saw them to call them "bobo," which means "stupid" in Spanish. In fact, they show off their blue feet in a dance to attract possible mates, and the bluest feet win the day. They gather in large colonies on land to sleep and mate, feeding at sea during the day. Their nostrils are permanently sealed shut, which allows them to plunge dive for fish.

| | |
|---|---|
| **SCIENTIFIC NAME** | *Sula nebouxii* |
| **ANIMAL GROUP** | Birds |
| **WINGSPAN/WEIGHT** | up to 5 ft./3.3 lb. |
| **DIET** | carnivore: fish |
| **LOCATION** | eastern Pacific coastlines |
| **STATUS** | Least Concern |

• June 26th •

# GILA MONSTER

This solitary venomous lizard is covered all over except on its underside with beadlike scales called osteoderms. Spending most of its time underground or under rocks, it comes out to warm up in the sun, like other reptiles, and hunt for food. It stores fat in its tail so that, in winter, it can go into brumation, which is what hibernation for cold-blooded animals is called (*see pp. 174–175*).

| | |
|---|---|
| **SCIENTIFIC NAME** | *Heloderma suspectum* |
| **ANIMAL GROUP** | Reptiles |
| **LENGTH/WEIGHT** | up to 1.8 ft. including tail/5 lb. |
| **DIET** | carnivore: small mammals, birds, reptiles, eggs |
| **LOCATION** | southwestern North America |
| **STATUS** | Near Threatened |

• June 27th •

# LONG-NOSED HORNED FROG

A ferocious predator that will even tackle scorpions because its prey do not see it coming, this amphibian looks amazingly like the dead leaves on the rainforest floor where it lives. It has a large head, a wide mouth and sounds more like a toad than a frog.

| | |
|---|---|
| **SCIENTIFIC NAME** | *Pelobatrachus nasuta* |
| **ANIMAL GROUP** | Amphibians |
| **LENGTH** | up to 5 in. |
| **DIET** | carnivore: arachnids, rodents, lizards, crabs, other frogs |
| **LOCATION** | southeastern Asia |
| **STATUS** | Least Concern |

• June 28th •

# SHORT-HORNED GRASSHOPPER

Grasshoppers hear each other's songs with tympana – membranes on either side of their abdomens under the wings that vibrate in response to sound waves. To produce their distinctive songs they stridulate, rubbing their hind legs against their forewings.

| | |
|---|---|
| **SCIENTIFIC NAME** | *Poecilotettix sanguineus* |
| **ANIMAL GROUP** | Invertebrates |
| **LENGTH** | up to 1.2 in. |
| **DIET** | herbivore: grasses, leaves |
| **LOCATION** | North America |
| **STATUS** | Not Evaluated |

• June 29th •

# CLOWNFISH

Also known as the anemonefish, this fish lives on coral reefs, where it has a symbiotic (cooperative) relationship with the sea anemone. The little fish is covered with a protective mucus that allows it to live and nest in the venomous tentacles of the anemone. There it is protected from predators. In return, the clownfish eats parasites and fans the saltwater with its tail to improve water circulation, which the anemone needs to get oxygen from the water.

| | |
|---|---|
| **SCIENTIFIC NAME** | *Amphiprion ocellaris* |
| **ANIMAL GROUP** | Fish |
| **LENGTH/WEIGHT** | up to 4.3 in./8 oz. |
| **DIET** | omnivore: algae, plankton, polychaete worms |
| **LOCATION** | Indian and Pacific Oceans |
| **STATUS** | Least Concern |

• June 30th •

# ELEPHANT SHREW

Also called the black and rufous sengi, this mammal gets its name from its long snout, which it uses to dig beetles and centipedes out of the soil to eat. It has a large territory and makes nests on the ground from the forest and woodland leaf litter. It can leap up to three feet in the air and run fast to escape a predator.

| | |
|---|---|
| **SCIENTIFIC NAME** | *Rhynchocyon petersi* |
| **ANIMAL GROUP** | Mammals |
| **LENGTH/WEIGHT** | up to 1.8 ft including tail/1.5 lb. |
| **DIET** | carnivore: insects |
| **LOCATION** | eastern Africa |
| **STATUS** | Least Concern |

# JULY

• July 1st •

## ESPAÑOLA GIANT TORTOISE

One of the largest tortoises in the world lives on the island of Española off the coast of Ecuador. It spends up to 16 hours a day resting under overhanging rocks, wallowing in puddles or warming up in the sun. The rest of the time it forages for grasses, cactus and fruit. Its long neck and limbs help it reach up into trees for food.

| | |
|---|---|
| SCIENTIFIC NAME | *Chelonoidis hoodensis* |
| ANIMAL GROUP | Reptiles |
| LENGTH/WEIGHT | up to 3 ft./155 lb. |
| DIET | herbivore: grasses, fruit |
| LOCATION | South America |
| STATUS | Critically Endangered |

• July 2nd •

# SCARLET MACAW

Flitting through the canopy of the rainforest, this very social parrot lives in a family group that may number up to 50 birds. It mates for life and nests high up in the trees, with both birds looking after the eggs and hatchlings. Macaws can eat unripe fruit that is toxic for other birds. They also cling in numbers to "clay licks," steep banks where scientists believe they supplement their diet with sodium-rich clay.

| | |
|---|---|
| **SCIENTIFIC NAME** | *Ara macao* |
| **ANIMAL GROUP** | Birds |
| **WINGSPAN/WEIGHT** | up to 3.5 ft./2.5 lb. |
| **DIET** | omnivore: nuts, fruit, insects, nectar |
| **LOCATION** | Central and South America |
| **STATUS** | Least Concern |

• July 3rd •

# EUROPEAN RABBIT

Able to turn their ears 270 degrees, run fast and jump over three feet off the ground, rabbits are constantly on the alert because they are a favorite prey for many other animals. They are found anywhere they can burrow, and their warrens can be very large, with up to 30 rabbits with their kits.

| | |
|---|---|
| **SCIENTIFIC NAME** | *Oryctolagus cuniculus* |
| **ANIMAL GROUP** | Mammals |
| **LENGTH/WEIGHT** | up to 1.3 ft./5 lb. |
| **DIET** | herbivore: grasses, herbs, roots, seeds, bark |
| **LOCATION** | Europe |
| **STATUS** | Endangered |

• July 4th •

# EUROPEAN MOLE

Constantly extending its tunnel system under farmland, grassland and gardens, the presence of this velvety short-sighted mammal is clear from above because it creates molehills by pushing loosened soil up shafts to the surface. It needs to eat at least 2 ounces of worms a day and takes live worms to store for later in a "larder."

| | |
|---|---|
| **SCIENTIFIC NAME** | *Talpa europaea* |
| **ANIMAL GROUP** | Mammals |
| **LENGTH/WEIGHT** | up to 9 in. including tail/4.5 oz. |
| **DIET** | carnivore: worms, insect larvae |
| **LOCATION** | Europe |
| **STATUS** | Least Concern |

• July 5th •

# PANTHER CHAMELEON

Found only on the island of Madagascar, this colorful chameleon lives in the trees of forests near rivers. It has a long, sticky tongue to lap up insects and eyes that can be moved independently of each other. To help it grip branches as it climbs, its five toes on each foot are fused together into two groups, one with two toes and the other with three. It can change color rapidly to attract a mate or confront a rival.

| | |
|---|---|
| **SCIENTIFIC NAME** | *Furcifer pardalis* |
| **ANIMAL GROUP** | Reptiles |
| **LENGTH/WEIGHT** | up to 1.6 ft./2.8 oz. |
| **DIET** | omnivore: insects, plants for hydration |
| **LOCATION** | Africa |
| **STATUS** | Least Concern |

• July 6th •

# FRILL-NECKED LIZARD

This dragonlike lizard with a tail that is around two-thirds of its body length is well camouflaged in the dry forests where it lives. It is an ambush predator that spends most of its time up in the trees but comes down to the ground after rain. If threatened, it raises its frill, but it can also move at up to 15 mph to get away or chase prey.

| | |
|---|---|
| **SCIENTIFIC NAME** | *Chlamydosaurus kingii* |
| **ANIMAL GROUP** | Reptiles |
| **LENGTH/WEIGHT** | up to 3.6 ft. including tail/28 oz. |
| **DIET** | carnivore: insects, caterpillars, small mammals |
| **LOCATION** | Australia |
| **STATUS** | Least Concern |

• July 7th •

# SUN BEAR

With its short black fur, round ears and bowed front legs, this smallest of all the bears is not a "sun" bear at all! Its name comes from the glowing patch of fur on its chest. Its long, curved claws can dig into the ground, open nests and climb trees. It has an excellent sense of smell which helps it find food. It searches at sunset and into the night, extracting honey from stingless bee nests and termites from mounds with its 10-inch-long tongue. Unlike most bears, the sun bear does not need to hibernate as it can find food all year.

| | |
|---|---|
| **SCIENTIFIC NAME** | *Helarctos malayanus* |
| **ANIMAL GROUP** | Mammals |
| **LENGTH/WEIGHT** | up to 5 ft. including tail/200 lb. |
| **DIET** | omnivore: termites, ants, bee larvae, honey |
| **LOCATION** | southeastern Asia |
| **STATUS** | Vulnerable |

• July 8th •

# NORTHERN CARDINAL

The male's tufted head and bright color are easily spotted, particularly against a snowy landscape. The more berries rich in red pigments it eats, the more red its feathers become. Both male and female cardinals have a repertoire of many different songs and whistles. The males fiercely defend breeding territories and have even been known to attack their own reflections in glass!

| | |
|---|---|
| **SCIENTIFIC NAME** | *Cardinalis cardinalis* |
| **ANIMAL GROUP** | Birds |
| **WINGSPAN/WEIGHT** | up to 1 ft./1.8 oz. |
| **DIET** | omnivore: seeds, insects, fruit |
| **LOCATION** | North America |
| **STATUS** | Least Concern |

# JAWS AND TEETH

There are some animals that have very powerful jaws and teeth and this can tell us a lot about the way they live. Most use these weapons every day when they are hunting, but others employ them for defense when threatened.

• July 9th •

## JAGUAR

With a bite more powerful than any other big cat (about twice that of a tiger), a jaguar can break through the shells of turtles and the armored skins of caimans.

| | |
|---|---|
| **SCIENTIFIC NAME** | *Panthera onca* |
| **ANIMAL GROUP** | Mammals |
| **LENGTH/WEIGHT** | up to 9 ft. including tail/265 lb. |
| **DIET** | carnivore: caimans, monkeys, fish, birds |
| **LOCATION** | Central and South America |
| **STATUS** | Near Threatened |

• July 10th •

## BURMESE PYTHON

One of the largest snakes in the world, this python has an amazing gape and is able to capture prey that is up to four times as wide as its head. It grasps its prey – anything from rats to alligators – with up to 120 teeth in two rows on the top jaw and one on the bottom. It then coils around the prey, constricting it tightly before swallowing it whole.

| | |
|---|---|
| **SCIENTIFIC NAME** | *Python bivittatus* |
| **ANIMAL GROUP** | Reptiles |
| **LENGTH/WEIGHT** | up to 23 ft./245 lb. |
| **DIET** | carnivore: mamma birds, reptiles |
| **LOCATION** | southeastern Asia |
| **STATUS** | Vulnerable |

• July 11th •

## GOBLIN SHARK

This deep-sea fish slingshot-feeds by extending its jaw to the same length as its nose to snatch prey. The jaw's ligaments allow it to do this and its mouth opens very wide. Sensing organs called ampullae of Lorenzini help it find its prey in the dark.

| | |
|---|---|
| **SCIENTIFIC NAME** | *Mitsukurina owstoni* |
| **ANIMAL GROUP** | Fish |
| **LENGTH/WEIGHT** | up to 20 ft./460 lb. |
| **DIET** | carnivore: dragonfish, squid, crustacear |
| **LOCATION** | Atlantic, Pacific and Indian Oceans |
| **STATUS** | Least Concern |

• July 12th •
# GULPER EEL

This is a bizarre-looking creature. Most of its length is its long, whiplike tail, but its enormous mouth is one-third of its body and can be opened very wide indeed. Its stomach can expand if it swallows large fish.

| | |
|---|---|
| SCIENTIFIC NAME | *Saccopharynx ampullaceus* |
| ANIMAL GROUP | Fish |
| LENGTH/WEIGHT | up to 5 ft./45 lb. |
| DIET | carnivore: small fish, squid, shrimps |
| LOCATION | Atlantic Ocean |
| STATUS | Least Concern |

• July 13th •
# SPOTTED HYENA

A fierce hunter as well as a scavenger, the hyena has a bite stronger in relation to its size than most other animals. Its jaws and teeth are very strong and it eats all of its prey, including bones, except horns.

| | |
|---|---|
| SCIENTIFIC NAME | *Crocuta crocuta* |
| ANIMAL GROUP | Mammals |
| LENGTH/WEIGHT | up to 6 ft. including tail/175 lb. |
| DIET | carnivore: zebra, gazelles, wildebeest, buffalo, giraffes |
| LOCATION | Africa |
| STATUS | Least Concern |

• July 14th •
# HIPPOPOTAMUS

In theory, the bite force of this animal is enough to bisect a crocodile (cut it in half), although as a herbivore, it would only attack such an animal if threatened. It has very sharp teeth and is very dangerous because of the size and weight that it can bring to bear.

| | |
|---|---|
| SCIENTIFIC NAME | *Hippopotamus amphibius* |
| ANIMAL GROUP | Mammals |
| LENGTH/WEIGHT | up to 18 ft. including tail/4.4 tons |
| DIET | herbivore: grasses, fruit, aquatic plants |
| LOCATION | Africa |
| STATUS | Vulnerable |

• July 15th •

# AFRICAN SAVANNA ELEPHANT

These are the largest living land animals on Earth. They live in family groups of females and their calves – up to 70 in total – led by a dominant female matriarch. The young are raised by the group and do not become independent until they are about eight years old. The bulls (males) live with a herd if they are young or are needed for mating. Elephants need to eat a lot every day, and when it is hot, often travel as much as 25 miles in a day searching for grass, trees and water.

| | |
|---|---|
| **SCIENTIFIC NAME** | *Loxodonta africana* |
| **ANIMAL GROUP** | Mammals |
| **LENGTH/WEIGHT** | up to 24 ft. including tail/8 tons |
| **DIET** | herbivore: grasses, leaves, bark, fruit |
| **LOCATION** | Africa |
| **STATUS** | Endangered |

• July 16th •

# NORTHERN GIRAFFE

The giraffe is a record breaker as well – it is the world's tallest mammal. Its incredibly long neck allows it to browse on trees that other animals cannot reach. The neck has the same number of vertebrae – seven – as humans, but each one is 10 inches long. Giraffes particularly like the leaves and shoots of the thorny acacia tree; the tough tissue of their tongues and mouths protects them from the thorns. Giraffes live in small family groups of females and young called "towers."

| | |
|---|---|
| **SCIENTIFIC NAME** | *Giraffa camelopardalis* |
| **ANIMAL GROUP** | Mammals |
| **LENGTH/WEIGHT** | up to 19 ft. including tail/2 tons |
| **DIET** | herbivore: leaves, flowers, fruit |
| **LOCATION** | Africa |
| **STATUS** | Vulnerable |

### • July 17th •

# CANE TOAD

Predators beware! When threatened, this toad's toxic skin exudes a milky-white fluid that can kill. Even its tadpoles will poison most animals if they are eaten. The toad uses sight and smell to find prey and eats a wide range of animals. It is called a "cane" toad because farmers used it to get rid of pests such as the cane beetle from crops of sugar cane.

| | |
|---|---|
| **SCIENTIFIC NAME** | *Rhinella marina* |
| **ANIMAL GROUP** | Amphibians |
| **LENGTH/WEIGHT** | up to 10 in./3.3 lb. |
| **DIET** | omnivore: small mammals, birds, invertebrates, plants |
| **LOCATION** | Central and South America |
| **STATUS** | Least Concern |

### • July 18th •

# GIRAFFE WEEVIL

Native to the island of Madagascar, this beetle has an extremely long neck that reminds people of a giraffe's. The male's neck is two to three times the length of the female's. Its bright red elytra, or forewings, cover and protect fragile flying wings. These weevils live and feed on a particular species of tree. After mating, the male rolls a leaf from this tree into a tube in which the female lays a single egg.

| | |
|---|---|
| **SCIENTIFIC NAME** | *Trachelophorus giraffa* |
| **ANIMAL GROUP** | Invertebrates |
| **BODY LENGTH** | up to 1 in. |
| **DIET** | herbivore: leaves |
| **LOCATION** | Africa |
| **STATUS** | Near Threatened |

• July 19th •

## NEON TETRA

In the warm waters of the Amazon basin, an unusual fish moves in very large shoals. It changes color as it goes, depending on the lighting of its environment. It becomes dull in darker waters, and bright blue in sunny streams. The change of color camouflages it from predators and also protects it from ultraviolet radiation. This small fish can move quickly – up to 15 mph – when it needs to.

| | |
|---|---|
| **SCIENTIFIC NAME** | *Paracheirodon innesi* |
| **ANIMAL GROUP** | Fish |
| **LENGTH** | up to 1.5 in. |
| **DIET** | omnivore: worms, insect larvae, algae |
| **LOCATION** | South America |
| **STATUS** | Not Evaluated |

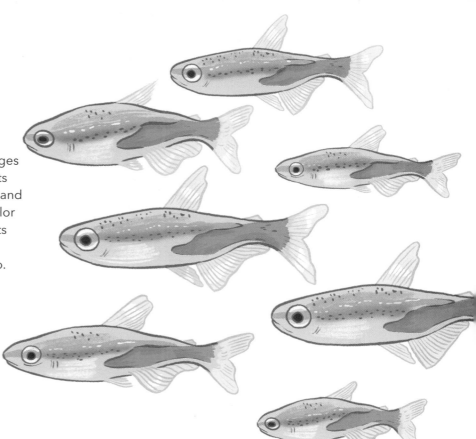

• July 20th •

## EUROPEAN HAMSTER

Also known as the black-bellied hamster, this small mammal lives on its own – except when there is a mother with young – in an underground burrow that is usually several yards long and up to 10 feet deep. The burrow has a living chamber, a food larder and toilet pits. The hamster hibernates (*see pp. 174–175*) during the winter, waking occasionally to feed on food it has stored.

| | |
|---|---|
| **SCIENTIFIC NAME** | *Cricetus cricetus* |
| **ANIMAL GROUP** | Mammals |
| **LENGTH/WEIGHT** | up to 15 in. including tail/1 lb. |
| **DIET** | omnivore: seeds, grasses, insects |
| **LOCATION** | Europe and western Asia |
| **STATUS** | Critically Endangered |

# FLIGHT

Wings enable birds, bats and insects to fly and glide, which most of them do with amazing ease. However, these two pages will also introduce you to some surprising gliding and flying wingless mammals, reptiles and fish.

### • July 21st •

## SOUTHERN FLYING SQUIRREL

This nocturnal squirrel can "fly" because a patagium, a furry flap of skin connecting wrists to ankles, acts as a parachute when the animal leaps from a tree and glides for up to 200 feet.

| | |
|---|---|
| SCIENTIFIC NAME | *Glaucomys volans* |
| ANIMAL GROUP | Mammals |
| LENGTH/WEIGHT | up to 15 in. including tail/3.2 oz. |
| DIET | omnivore: berries, seeds, bark, nuts, insects |
| LOCATION | southeastern North America |
| STATUS | Least Concern |

### • July 22nd •

## PARADISE TREE SNAKE

By flattening its body, this snake can glide from a tree down onto prey below. It undulates – moves in a wavelike pattern – to control its direction of flight.

| | |
|---|---|
| SCIENTIFIC NAME | *Chrysopelea paradisi* |
| ANIMAL GROUP | Reptiles |
| LENGTH/WEIGHT | up to 3 ft./2 lb. |
| DIET | carnivore: lizards, frogs, other small animals |
| LOCATION | southeastern Asia |
| STATUS | Least Concern |

### • July 23rd •

## FLYING DRAGON LIZARD

Found in tropical rainforests, this lizard has a set of long ribs joined by skin membranes that it can extend as wings to catch the wind and glide.

| | |
|---|---|
| SCIENTIFIC NAME | *Draco volans* |
| ANIMAL GROUP | Reptiles |
| LENGTH/WEIGHT | up to 8 in./3.9 oz. |
| DIET | carnivore: ants |
| LOCATION | southeastern Asia |
| STATUS | Least Concern |

• July 24th •

# GRAY LONG-EARED BAT

All bats, including this one, have great maneuverability because of their many-jointed wings, and are arguably more efficient than birds in the air. When bats hunt at night, they use echolocation to find prey. By listening to echoes, they navigate and find insects.

| | |
|---|---|
| **SCIENTIFIC NAME** | *Plecotus austriacus* |
| **ANIMAL GROUP** | Mammals |
| **WINGSPAN/WEIGHT** | up to 1 ft./0.4 oz. |
| **DIET** | carnivore: moths, beetles, flying ants |
| **LOCATION** | Europe |
| **STATUS** | Near Threatened |

• July 25th •

# ANDEAN CONDOR

High above the Andes Mountains at heights of up to 3.1 miles above sea level, this bird has the longest wingspan in the world. It can stay aloft for more than 90 miles without flapping its wings once.

| | |
|---|---|
| **SCIENTIFIC NAME** | *Vultur gryphus* |
| **ANIMAL GROUP** | Birds |
| **WINGSPAN/WEIGHT** | up to 11 ft./33 lb. |
| **DIET** | carnivore: carrion |
| **LOCATION** | western South America |
| **STATUS** | Vulnerable |

• July 26th •

# BENNETT'S FLYING FISH

In order to escape predators in the water, this torpedo-shaped fish can "fly" nearly 10 feet above the surface of the sea for 650 feet at a time. It exits the water at speeds of more than 30 mph, then flattens its fins against its body. However, it may fall prey to another predator – a bird!

| | |
|---|---|
| **SCIENTIFIC NAME** | *Cheilopogon pinnatibarbatus* |
| **ANIMAL GROUP** | Fish |
| **LENGTH/WEIGHT** | up to 20 in./2.2 oz. |
| **DIET** | omnivore: plankton, crustaceans |
| **LOCATION** | southeastern Pacific Ocean |
| **STATUS** | Least Concern |

• July 27th •

## VIETNAMESE MOSSY FROG

Also called the Tonkin bug-eyed frog, this semiaquatic amphibian shelters in water-filled tree holes or in wet moss with only its eyes showing as it keeps watch. It folds into a ball and plays dead when threatened by a snake or other predator.

| | |
|---|---|
| **SCIENTIFIC NAME** | *Theloderma corticale* |
| **ANIMAL GROUP** | Amphibians |
| **LENGTH** | up to 3 in. |
| **DIET** | carnivore: crickets, cockroaches |
| **LOCATION** | eastern Asia |
| **STATUS** | Least Concern |

• July 28th •

## SEVEN-SPOT LADYBIRD

The seven symmetrical black spots on its two red wing cases make this insect easy to see. Ladybirds hibernate in winter, emerging in spring to look for food. Gardeners like them because they eat up to 5,000 aphids in their year-long lives.

| | |
|---|---|
| **SCIENTIFIC NAME** | *Coccinella septempunctata* |
| **ANIMAL GROUP** | Invertebrates |
| **LENGTH** | up to 0.3 in. |
| **DIET** | carnivore: aphids, other insects |
| **LOCATION** | Europe, Asia, North America |
| **STATUS** | Not Evaluated |

• July 29th •

## EASTERN MEADOWLARK

This bird finds a fencepost or other perch and then sings its heart out with whistling songs. It walks when it is on the ground and, in the air, flutters rapidly with short glides. In a flock, it hunts insects in the fields.

| | |
|---|---|
| **SCIENTIFIC NAME** | *Sturnella magna* |
| **ANIMAL GROUP** | Birds |
| **WINGSPAN/WEIGHT** | up to 16 in./5 oz. |
| **DIET** | omnivore: insects, fruit, seeds, caterpillars |
| **LOCATION** | North America, northern South America |
| **STATUS** | Near Threatened |

• July 30th •
# DUCK-BILLED PLATYPUS

With thick fur to repel water and keep them warm and dry, platypuses can stay underwater for up to two minutes. Sealing their nostrils shut, they send out electrical impulses to find prey. Receptors on their soft, duckbill-like snouts tell them where to look. These mammals are monotremes, laying eggs instead of giving birth to live young.

| | |
|---|---|
| **SCIENTIFIC NAME** | *Ornithorhynchus anatinus* |
| **ANIMAL GROUP** | Mammals |
| **LENGTH/WEIGHT** | up to 2 ft. including tail/6 lb. |
| **DIET** | carnivore: worms, larvae, shellfish |
| **LOCATION** | eastern Australia |
| **STATUS** | Near Threatened |

• July 31st •
# PALMATE NEWT

A freshwater amphibian, this newt can be seen in shallow ponds on marshes, heathlands and bogs. It spends most of the day in thick aquatic vegetation, emerging into open water after dark. It is often found on land under logs or other debris.

| | |
|---|---|
| **SCIENTIFIC NAME** | *Lissotriton helveticus* |
| **ANIMAL GROUP** | Amphibian |
| **LENGTH/WEIGHT** | up to 3.7 in./2.1 oz. |
| **DIET** | carnivore: invertebrates, tadpoles, other newts |
| **LOCATION** | western Europe |
| **STATUS** | Least Concern |

# AUGUST

• August 1st •

## RED-CROWNED CRANE

Spreading their wings and calling to the sky, a pair of cranes dance in unison. This courtship ritual strengthens their lifelong relationship. Both birds will then build a nest and share the incubation of the eggs. They feed in deep-water marshes, probing the water with their long bills as they walk.

| | |
|---|---|
| **SCIENTIFIC NAME** | *Grus japonensis* |
| **ANIMAL GROUP** | Birds |
| **WINGSPAN/WEIGHT** | up to 8 ft./22 lb. |
| **DIET** | omnivore: crabs, fish, worms, grasses, reeds |
| **LOCATION** | eastern Asia |
| **STATUS** | Vulnerable |

• August 2nd •

# BLACK-CROWNED SQUIRREL MONKEY

Squirrel monkeys are named for the quick, agile way they climb and leap through the trees. The black-crowned squirrel monkeys look for food at all levels of the rainforest. When on the ground, they walk on all fours and run. Groups may number up to 100, but they forage in smaller "troops." The young travel at first on their mothers' backs, but then are looked after by "aunts" in the group.

| | |
|---|---|
| **SCIENTIFIC NAME** | *Saimiri oerstedii* |
| **ANIMAL GROUP** | Mammals |
| **LENGTH/WEIGHT** | up to 2.5 ft. including tail/2.1 lb. |
| **DIET** | omnivore: insects, fruits, small mammals including bats |
| **LOCATION** | western Central America |
| **STATUS** | Endangered |

• August 3rd •

# GIANT DAY GECKO

A native of the island of Madagascar, this lizard is usually seen on large trees in the forest or clinging to the walls of buildings in towns. It does not have claws, but the lamellae – thin scales covered in hairlike setae – on its feet allow it to climb smooth surfaces. It has no eyelids, so licks its eyes with its long tongue to keep them moist and clean. It is very territorial and will only let females into its area.

| | |
|---|---|
| **SCIENTIFIC NAME** | *Phelsuma grandis* |
| **ANIMAL GROUP** | Reptiles |
| **LENGTH/WEIGHT** | up to 12 in. including tail/2.5 oz. |
| **DIET** | omnivore: insects, crabs, nectar |
| **LOCATION** | Africa |
| **STATUS** | Least Concern |

• August 4th •

## BLUE MOUNTAIN SWALLOWTAIL

Also known as the blue emperor, this very beautiful butterfly is found in the canopy of tropical rainforests. It has a short life, from as little as a week to eight months, and will migrate if it needs to find warmth and food.

| | |
|---|---|
| **SCIENTIFIC NAME** | *Papilio ulysses* |
| **ANIMAL GROUP** | Invertebrates |
| **WINGSPAN** | up to 5 in. |
| **DIET** | herbivore: flowers, nectar |
| **LOCATION** | Australia, Asia |
| **STATUS** | Not Evaluated |

• August 5th •

## LAUGHING KOOKABURRA

This characterful kingfisher lives in dense woodland near water, nesting 30 feet up in mountain gum trees. Its unique song in the early morning and early evening all year sounds much like a human laughing. It has other songs and calls to talk to birds in its small family group.

| | |
|---|---|
| **SCIENTIFIC NAME** | *Dacelo novaeguineae* |
| **ANIMAL GROUP** | Birds |
| **WINGSPAN/WEIGHT** | up to 2.1 ft./1 lb. |
| **DIET** | carnivore: insects, worms, crustaceans, fish, frogs |
| **LOCATION** | eastern Australia |
| **STATUS** | Least Concern |

• August 6th •

## GIANT CLAM

This is the largest mollusk on Earth. It is a bivalve – it has a shell with two halves – that lives in the warm waters of coral reefs. When it is fully grown, it cannot close its shell. It filter-feeds on microscopic plants and animals, and other nutrients from saltwater.

| | |
|---|---|
| **SCIENTIFIC NAME** | *Tridacna gigas* |
| **ANIMAL GROUP** | Invertebrates |
| **LENGTH/WEIGHT** | up to 5 ft./550 lb. |
| **DIET** | omnivore: zooplankton, phytoplankton |
| **LOCATION** | Indian and Pacific Oceans |
| **STATUS** | Vulnerable |

• August 7th •

## GREAT BUSTARD

The male bustard is up to 50 percent bigger than the female and sports beardlike white feathers that may be nearly 8 inches long. These ground-living birds live on grasslands in groups called "droves" that wander together in search of food. Males and females live in separate droves except when mating in spring. The males compete at an area called a "lek," where they perform flamboyant displays to impress the females.

| | |
|---|---|
| **SCIENTIFIC NAME** | *Otis tarda* |
| **ANIMAL GROUP** | Birds |
| **WINGSPAN/WEIGHT** | up to 9 ft./45 lb. |
| **DIET** | omnivore: shoots, leaves, berries, flowers, insects |
| **LOCATION** | Asia, southern Europe |
| **STATUS** | Vulnerable |

• August 8th •

## BLACK RHINOCEROS

Living on the savanna and in shrublands, this large, heavy animal forages for food and drink when it is cool or at night. It mostly browses bushes or trees rather than grazing on grass. During the heat of the day, it snoozes in the shade or wallows in shallow mud pools. Its horns are used to defend its territory and calves, dig for water and break branches. They are made of keratin, like human hair and nails, and grow continually throughout its life.

| | |
|---|---|
| **SCIENTIFIC NAME** | *Diceros bicornis* |
| **ANIMAL GROUP** | Mammals |
| **LENGTH/WEIGHT** | up to 15 ft. including tail/1.5 tons |
| **DIET** | herbivore: woody plants, leaves |
| **LOCATION** | Africa |
| **STATUS** | Critically Endangered |

• August 9th •

# AMAZON LEAF FISH

A true master of camouflage, this fish looks just like a fallen leaf underwater and can change from browns to yellows to match its surroundings. It even copies the way that leaves drift in shallow water, so its prey do not see it coming. It has a massive mouth and strikes with lightning speed of 0.2 seconds or less to grab prey.

| | |
|---|---|
| **SCIENTIFIC NAME** | *Monocirrhus polyacanthus* |
| **ANIMAL GROUP** | Fish |
| **LENGTH** | up to 4 in. |
| **DIET** | carnivore: fish, insects |
| **LOCATION** | South America |
| **STATUS** | Not Evaluated |

# TREE LIVING

Trees provide food and shelter for many different animals. But life in the trees is not without challenges and animals have found many different ways to adapt, either physically or behaviorally.

## • August 10th •
### TREECREEPER

This little bird literally creeps up tree trunks to forage for insects or to roost and nest safely. It does not creep down; rather it flutters to the ground then creeps up again. Its bright white chin reflects light onto the wood and this helps it spot its prey in crevices in the bark.

| | |
|---|---|
| SCIENTIFIC NAME | *Certhia familiaris* |
| ANIMAL GROUP | Birds |
| WINGSPAN/WEIGHT | up to 8 in./0.4 oz. |
| DIET | omnivore: insects, spiders, seeds |
| LOCATION | Europe, Asia |
| STATUS | Least Concern |

## • August 11th •
### FLAT BARK BEETLE

The shape of these beetles is an example of how animals adapt to the place they live in. The flatness of their bodies allows them to crawl under the bark of decaying dead trees and find the fungi and larvae they need to survive.

| | |
|---|---|
| SCIENTIFIC NAME | *Cucujus cinnaberinus* |
| ANIMAL GROUP | Invertebrates |
| LENGTH | up to 0.6 in. |
| DIET | omnivore: tree bark, insect larvae, fungi |
| LOCATION | Europe |
| STATUS | Near Threatened |

## • August 12th •
### RED SQUIRREL

An agile climber that is able to jump more than 6 feet from branch to branch, this squirrel has double-jointed ankles that allow it to go down trees headfirst. Its young are born high up in the trees, in nests called dreys that are made from twigs lined with moss, grass and leaves. It stores food for winter because it does not hibernate (see pp.174–175).

| | |
|---|---|
| SCIENTIFIC NAME | *Sciurus vulgaris* |
| ANIMAL GROUP | Mammals |
| LENGTH/WEIGHT | up to 17 in. including tail/12 oz. |
| DIET | omnivore: seeds, fungi, bark, eggs |
| LOCATION | Europe, Asia |
| STATUS | Least Concern |

## • August 13th •
## HAINAN GIBBON

This gibbon is only found on Hainan Island off the coast of China. Its hooked hands allow it to hang suspended for a long time and brachiate very easily from branch to branch, reaching up to 39 feet in one swing at speeds of up to 35 mph.

| | |
|---|---|
| SCIENTIFIC NAME | *Nomascus hainanus* |
| ANIMAL GROUP | Mammals |
| LENGTH/WEIGHT | up to 19 in./22 lb. |
| DIET | omnivore: leaves, shoots, fruit, eggs, insects |
| LOCATION | Asia |
| STATUS | Critically Endangered |

## • August 14th •
## EMERALD TREE MONITOR

In the rainforests of Papua New Guinea lives a slender monitor lizard that uses its tail as a fifth limb. It wraps it around branches and uses its specially adapted feet to move through the trees. Small groups with a dominant male forage and rest together.

| | |
|---|---|
| SCIENTIFIC NAME | *Varanus prasinus* |
| ANIMAL GROUP | Reptiles |
| LENGTH/WEIGHT | up to 3 ft. including tail/10 oz. |
| DIET | carnivore: insects, frogs, lizards, eggs, small mammals |
| LOCATION | Asia |
| STATUS | Least Concern |

## • August 15th •
## KINKAJOU

Also known as the honey bear because of its color, this rainforest mammal spends most of its time in the trees. It is nocturnal, often hanging upside down by its long tail to feed. During the day, it sleeps in tree holes.

| | |
|---|---|
| SCIENTIFIC NAME | *Potos flavus* |
| ANIMAL GROUP | Mammals |
| LENGTH/WEIGHT | up to 4.2 ft. including tail/10 lb. |
| DIET | omnivore: fruits, insects, nectar |
| LOCATION | Central America, northern South America |
| STATUS | Least Concern |

• August 16th •
# SUMATRAN ORANGUTAN

The island of Sumatra is the only place in the world where tigers, orangutans and elephants (see p.209) live together. The orangutan is almost entirely arboreal, brachiating from branch to branch on its long arms – it measures around 7 feet from fingertip to fingertip. It makes sleeping nests every night by pulling branches together and weaving in twigs and leaves.

**SCIENTIFIC NAME** *Pongo abelii*
**ANIMAL GROUP** Mammals
**HEIGHT/WEIGHT** up to 5 ft./195 lb.
**DIET** omnivore: fruit, leaves, seeds, insects
**LOCATION** southeastern Asia
**STATUS** Critically Endangered

• August 17th •

# SUMATRAN TIGER

The smallest member of its family, this tiger lives only on the island of Sumatra. It has the narrowest stripe pattern of all the tigers, which helps to camouflage it well in the dense tropical forest. Each tiger's stripe pattern is unique to that tiger. The Sumatran tiger hunts wild pigs and deer at night and can run after its prey at up to 60 mph.

| | |
|---|---|
| SCIENTIFIC NAME | *Panthera tigris* |
| ANIMAL GROUP | Mammals |
| LENGTH/WEIGHT | up to 11 ft. including tail/310 lb. |
| DIET | carnivore: deer, wild boar, macaque monkeys |
| LOCATION | southeastern Asia |
| STATUS | Critically Endangered |

• August 18th •

# MOUNTAIN GAZELLE

Living in small groups of up to eight, this browser gets liquid from its food if there is a drought. It is one of the few mammals in which both sexes have horns. Those of the male are wider and longer and used to fight off rivals. Its slender build and long legs enable it to run away at up to 50 mph when threatened by a predator.

| | |
|---|---|
| SCIENTIFIC NAME | *Gazella gazella* |
| ANIMAL GROUP | Mammals |
| LENGTH/WEIGHT | up to 4.3 ft. including tail/65 lb. |
| DIET | herbivore: leaves, grasses, shrubs |
| LOCATION | Africa, Asia |
| STATUS | Endangered |

• August 19th •

# EUROPEAN GREEN TOAD

Found in all kinds of habitats, ranging from high plains and mountains to semi-deserts and urban areas, this toad is always near water. During the day, it shelters in its den, normally a burrow underground, and to keep cool, it sits in shallow water. It regularly sheds its skin and also changes color in response to changes in temperature or light.

| | |
|---|---|
| SCIENTIFIC NAME | *Bufotes viridis* |
| ANIMAL GROUP | Amphibians |
| LENGTH/WEIGHT | up to 4.7 in./3.5 oz. |
| DIET | carnivore: invertebrates |
| LOCATION | western Europe |
| STATUS | Least Concern |

• August 20th •
# RED LIONFISH

Moving slowly or staying very still in the warm, tropical waters, this ambush predator waits for small fish to get close enough. Then it stretches out its winglike pectoral fins to trap the prey before swallowing it whole. The venomous spines along its back are used to defend itself when attacked by a predator such as the coastal shark. The females can release around two million eggs a year!

| | |
|---|---|
| SCIENTIFIC NAME | *Pterois volitans* |
| ANIMAL GROUP | Fish |
| LENGTH/WEIGHT | up to 15 in./2.6 lb. |
| DIET | carnivore: crustaceans, fish |
| LOCATION | Indian and Pacific Oceans |
| STATUS | Least Concern |

• August 21st •
# BROWN PELICAN

This smallest member of the pelican family is one of only two species of pelicans to plunge-dive into the sea to get food. It can spot a fish near the surface from 65 feet up in the air, and air sacs make it buoyant when it is in the water so it cannot dive deep. Its beak is 9 inches long, and it uses the throat pouch under this to scoop up the fish. It tips its beak downward to drain the water before swallowing its prey. Brown pelicans usually stay in flocks on the coast, where they nest, fish and fly.

| | |
|---|---|
| SCIENTIFIC NAME | *Pelecanus occidentalis* |
| ANIMAL GROUP | Birds |
| WINGSPAN/WEIGHT | up to 7 ft./11 lb. |
| DIET | carnivore: fish, sardines, shrimps, carrion |
| LOCATION | North, Central and South America |
| STATUS | Least Concern |

• August 22nd •

# AARDWOLF

Not related either to an aardvark or a wolf, this is an insect-eating hyena that lives on grassy plains or in woodlands. Its name means "earth-wolf." It digs underground tunnels to sleep in by day and emerges to find food when it gets dark. It can lap up 250,000 termites in a single night with its long, sticky tongue.

| | |
|---|---|
| **SCIENTIFIC NAME** | *Proteles cristatus* |
| **ANIMAL GROUP** | Mammals |
| **LENGTH/WEIGHT** | up to 3.6 ft. including tail/30 lb. |
| **DIET** | carnivore: termites, larvae, eggs |
| **LOCATION** | eastern and southern Africa |
| **STATUS** | Least Concern |

• August 23rd •

# AFRICAN HERMIT SPIDER

This spider is an orb-weaver, building a large circular web made of thin silk. The female spends its day on the intricate web, which is attached to tree trunks, rocks or under the overhang of roofs in tropical regions. The web has a funnel-shaped retreat for her on the side. The supporting lines are strong enough to hold struggling insects when they fly into the trap. The male spider is much smaller than the female.

| | |
|---|---|
| **SCIENTIFIC NAME** | *Nephilingis cruentata* |
| **ANIMAL GROUP** | Invertebrates |
| **LEGSPAN** | up to 2.8 in. (female) |
| **DIET** | carnivore: crickets, other insects, spiders |
| **LOCATION** | Africa |
| **STATUS** | Least Concern |

• August 24th •

## AMAZON TREE BOA

Moving from branch to branch, this snake does hunt on the ground, but spends most of its time high up in rainforest trees. It is nocturnal, waiting wrapped around a branch to ambush prey such as opossums, iguanas or bats. When the prey gets close enough, it swings down from the branch, anchored by its prehensile tail. It swallows the animal whole, retreats and spends the next week digesting slowly.

| | |
|---|---|
| **SCIENTIFIC NAME** | *Corallus hortulana* |
| **ANIMAL GROUP** | Reptiles |
| **LENGTH/WEIGHT** | up to 7 ft./ 2 lb. |
| **DIET** | carnivore: small mammals, lizards, birds |
| **LOCATION** | South America |
| **STATUS** | Least Concern |

• August 25th •

## COPPERBAND BUTTERFLYFISH

Its rounded, flat body resembles the wing of a butterfly as it swims in and out of the corals and seagrass beds of its habitat. The young fry swim in schools (groups), but adults normally travel either alone or in pairs. Its elongated snout is shaped to pry small invertebrates from narrow crevices in the reef, and the fake eyespots either side of its body help protect it from eels, sharks and other predators.

| | |
|---|---|
| **SCIENTIFIC NAME** | *Chelmon rostratus* |
| **ANIMAL GROUP** | Fish |
| **LENGTH** | up to 8 in. |
| **DIET** | carnivore: crustaceans, shrimps, tubeworms |
| **LOCATION** | Indian and Pacific Oceans |
| **STATUS** | Least Concern |

# ON THE CORAL REEF

Coral reefs are underwater structures built by tiny sea animals that are vital habitats for sea life. It is estimated that they support 25 percent of all marine life, including around 4,000 different known species of fish.

## • August 26th •
### GREEN SEA TURTLE

This turtle and reefs have a symbiotic relationship. The turtle acts as a "gardener," grazing on seagrass and sponges, stopping them from growing too much. In turn, the reef provides food and protection for the turtle.

| | |
|---|---|
| SCIENTIFIC NAME | *Chelonia mydas* |
| ANIMAL GROUP | Reptiles |
| LENGTH/WEIGHT | up to 4.6 ft./395 lb. |
| DIET | omnivore: algae, seagrasses, jellyfish |
| LOCATION | Indian, Pacific and Atlantic Oceans |
| STATUS | Endangered |

## • August 27th •
### ZEBRA SEAHORSE

In the warm waters off northern Australia, coral provides plenty of food and places for this tiny seahorse to hide. It anchors itself with its tail to the coral or seaweed at depths of up to 230 feet.

| | |
|---|---|
| SCIENTIFIC NAME | *Hippocampus zebra* |
| ANIMAL GROUP | Fish |
| LENGTH | up to 3.5 in. |
| DIET | carnivore: small crustaceans |
| LOCATION | Indian and Pacific Oceans |
| STATUS | Data Deficient |

## • August 28th •
### POTTER'S ANGELFISH

This brightly colored fish fits in well in the equally brightly colored habitat of the reef. It has comblike teeth to pull food off hard reef surfaces during the day, and hides in crevices at night.

| | |
|---|---|
| SCIENTIFIC NAME | *Centropyge potteri* |
| ANIMAL GROUP | Fish |
| LENGTH | up to 6 in. |
| DIET | omnivore: algae, marine invertebrates |
| LOCATION | Pacific Ocean |
| STATUS | Least Concern |

## • August 29th •
## WHALE SHARK

Coral reefs are the breeding grounds for plankton and krill, the favorite foods of this shark, the largest fish in the sea. Whale sharks travel slowly around tropical and temperate oceans. Many of these wide-mouthed fish come together to filter-feed near reefs, where the food is abundant.

| | |
|---|---|
| **SCIENTIFIC NAME** | *Rhincodon typus* |
| **ANIMAL GROUP** | Fish |
| **LENGTH/WEIGHT** | up to 45 ft./25 tons |
| **DIET** | omnivore: plankton, krill, algae, larvae |
| **LOCATION** | Atlantic, Indian and Pacific Oceans |
| **STATUS** | Endangered |

## • August 30th •
## INDO-PACIFIC BOTTLENOSE DOLPHIN

Scientists have studied groups of these dolphins rubbing their bodies against corals and teaching their young to do the same. They think the dolphins do this to release compounds that will cure their skin ailments.

| | |
|---|---|
| **SCIENTIFIC NAME** | *Tursiops aduncus* |
| **ANIMAL GROUP** | Mammals |
| **LENGTH/WEIGHT** | up to 9 ft./510 lb. |
| **DIET** | carnivore: fish, squid, octopus |
| **LOCATION** | Indian and Pacific Oceans |
| **STATUS** | Near Threatened |

## • August 31st •
## CARIBBEAN REEF OCTOPUS

With a mantle (body) length under 5 inches, this is a small octopus with long arms. It lives on coral reefs along the north coast of South America, emerging at night to find crustaceans on the sea floor. It deters predators by sending out clouds of dark ink.

| | |
|---|---|
| **SCIENTIFIC NAME** | *Octopus briareus* |
| **ANIMAL GROUP** | Invertebrates |
| **ARM LENGTH/WEIGHT** | up to 2 ft./3.3 lb. |
| **DIET** | carnivore: crabs, clams, lobsters, snails |
| **LOCATION** | Atlantic Ocean |
| **STATUS** | Least Concern |

# SEPTEMBER

• September 1ˢᵗ •

## SPINNER DOLPHIN

Famous for its acrobatics, this dolphin is aptly named, as it regularly leaps out of the water and twists its body in the air, spinning up to seven times before splashing back down. It may do this to get rid of parasites, to attract a mate or to communicate with its pod. A pod of dolphins hunts and feeds at night, often herding fish into dense groups to take them easily.

| | |
|---|---|
| **SCIENTIFIC NAME** | *Stenella longirostris* |
| **ANIMAL GROUP** | Mammals |
| **LENGTH/WEIGHT** | up to 7 ft./175 lb. |
| **DIET** | carnivore: fish, squid, shrimps |
| **LOCATION** | Pacific, Atlantic, Indian Oceans |
| **STATUS** | Least Concern |

# EURASIAN BADGER

Sniffing out prey on its own at night, this powerful animal lives in an underground sett that is home to up to eight badgers. It can dig out and eat more than 200 worms in a single night, and will eat anything it can find. It fiercely defends its territory and its loose-fitting skin makes it difficult for any challenger to get a grip.

| | |
|---|---|
| SCIENTIFIC NAME | *Meles meles* |
| ANIMAL GROUP | Mammals |
| LENGTH/WEIGHT | up to 4 ft. including tail/25 lb. |
| DIET | omnivore: worms, snails, fruit, plants |
| LOCATION | Europe, central Asia |
| STATUS | Least Concern |

• September 3rd •

## FALSE CORAL SNAKE

This snake is well protected because it looks like a venomous coral snake, so predators avoid it. It lives in the rainforest, hunting on the ground late or early in the day mainly for snakes, which it often eats tail first.

| | |
|---|---|
| **SCIENTIFIC NAME** | *Erythrolamprus aesculapii* |
| **ANIMAL GROUP** | Reptiles |
| **LENGTH/WEIGHT** | up to 5 ft./2 oz. |
| **DIET** | carnivore: snakes, fish, worms, lizards |
| **LOCATION** | South America |
| **STATUS** | Least Concern |

• September 4th •

## RESPLENDENT QUETZAL

In the canopy of cloud forests of Central America lives a small bird that has one of the most splendid tails in the bird world and iridescent feathers all over that shine when they catch the light. The tail is up to 3 feet long, nearly double the length of its body. It is the national bird of Guatemala.

| | |
|---|---|
| **SCIENTIFIC NAME** | *Pharomachrus mocinno* |
| **ANIMAL GROUP** | Birds |
| **WINGSPAN/WEIGHT** | up to 22 in./7 oz. |
| **DIET** | omnivore: mainly fruit; also insects, small frogs, lizards |
| **LOCATION** | Central America |
| **STATUS** | Near Threatened |

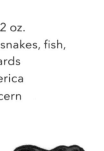

• September 5th •

## CENTRAL COAST STUBFOOT TOAD

Unusually, this toad has no external vocal sac, so to defend its territory or attract a mate, it can only produce calls that reach a distance of less than 25 feet. It is found near fast-flowing streams in lowland rainforests and lays its eggs in the water, where the hatched tadpoles cling to rocks.

| | |
|---|---|
| **SCIENTIFIC NAME** | *Atelopus franciscus* |
| **ANIMAL GROUP** | Amphibians |
| **LENGTH** | up to 1 in. |
| **DIET** | carnivore: insects, other invertebrates |
| **LOCATION** | northeastern South America |
| **STATUS** | Least Concern |

# ANIMAL ASSASSINS

To catch prey or defend themselves, animals have all kinds of strategies that serve them well, especially when accompanied by the element of surprise. Here are a few of the world's most deadly animal assassins.

• September 6th •

## HOODED PITOHUI

This bird of the tropical forests of New Guinea is poison to the touch! It absorbs one of the most toxic natural substances known from its diet of small melyrid beetles.

| | |
|---|---|
| **SCIENTIFIC NAME** | *Pitohui dichrous* |
| **ANIMAL GROUP** | Birds |
| **WINGSPAN/WEIGHT** | up to 9 in./2.7 oz. |
| **DIET** | omnivore: fruit, seeds, beetles, fli |
| **LOCATION** | Asia |
| **STATUS** | Least Concern |

• September 7th •

## YELLOW ASSASSIN FLY

Also called the robber fly, this predator mimics the bumblebee so that predators think it is armed with a sting. Instead, it pounces on unsuspecting smaller insects.

| | |
|---|---|
| **SCIENTIFIC NAME** | *Laphria flava* |
| **ANIMAL GROUP** | Invertebrates |
| **LENGTH** | up to 1 in. |
| **DIET** | carnivore: insects, larvae |
| **LOCATION** | Europe |
| **STATUS** | Not Evaluated |

• September 8th •

## GREATER BLUE-RINGED OCTOPUS

Beware this octopus – it is only as big as the length of a pencil but is one of the most toxic marine animals. It has two kinds of venom in its saliva. One is used to immobilize prey. The second is used for defense and can kill a human when bitten.

| | |
|---|---|
| **SCIENTIFIC NAME** | *Hapalochlaena lunulata* |
| **ANIMAL GROUP** | Invertebrates |
| **LENGTH/WEIGHT** | up to 6 in. including arms/3.5 oz. |
| **DIET** | carnivore: small crabs, shrimps, fish |
| **LOCATION** | Indian and Pacific Oceans |
| **STATUS** | Least Concern |

• September 9th •

# ALLIGATOR SNAPPING TURTLE

This turtle has simply to stay motionless, open its mouth and wiggle its pink, worm-shaped tongue to attract prey. Fish swim in and are "snapped" up. One of the largest freshwater turtles, it can stay submerged for up to 50 minutes.

| | |
|---|---|
| SCIENTIFIC NAME | *Macrochelys temminckii* |
| ANIMAL GROUP | Reptiles |
| LENGTH/WEIGHT | up to 3.3. ft. including tail/200 lb. |
| DIET | carnivore: fish, mollusks, turtles, insects |
| LOCATION | North America |
| STATUS | Vulnerable |

• September 10th •

# BLACK PANTHER

This melanistic (black coated), yellow-eyed jaguar simply fades into the shadows when it hunts at night. It walks softly, using stealth and speed to capture prey, which it dispatches with its strong jaws and sharp teeth.

| | |
|---|---|
| SCIENTIFIC NAME | *Panthera onca* |
| ANIMAL GROUP | Mammals |
| LENGTH/WEIGHT | up to 10 ft. including tail/350 lb. |
| DIET | carnivore: deer, wild hogs, raccoons |
| LOCATION | southern North America, Central and South America |
| STATUS | Near Threatened |

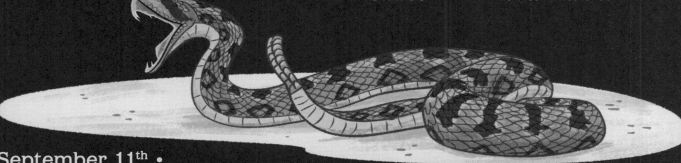

• September 11th •

# TIMBER RATTLESNAKE

Like all other pit vipers, this venomous snake has pit organs between its nostrils and eyes to sense changes in temperature, which helps when hunting warm-blooded rodents at night. It is very well camouflaged, so blends into its surroundings before striking and killing.

| | |
|---|---|
| SCIENTIFIC NAME | *Crotalus horridus* |
| ANIMAL GROUP | Reptiles |
| LENGTH/WEIGHT | up to 7 ft./2.2 oz. |
| DIET | carnivore: squirrels, rabbits, birds, lizards, frogs |
| LOCATION | eastern North America |
| STATUS | Least Concern |

• September 12th •
# FOUR-TOED HEDGEHOG

This prickly pygmy lives on the savanna and in the grasslands. When threatened, it rolls into a ball about the size of a grapefruit. It is nocturnal, relying on its strong sense of smell and excellent hearing to find food in the dark. It may hibernate or estivate (see pp. 174-175) if it gets too cold or hot. It nests among rocks or hollow trees and its young are called hoglets.

| | |
|---|---|
| **SCIENTIFIC NAME** | *Atelerix albiventris* |
| **ANIMAL GROUP** | Mammals |
| **LENGTH/WEIGHT** | up to 11 in./2.2 oz. |
| **DIET** | omnivore: insects, spiders, scorpions, seeds, roots |
| **LOCATION** | central Africa |
| **STATUS** | Least Concern |

• September 13th •
# TOWNSEND'S BIG-EARED BAT

Its large ears funnel sound and help regulate the bat's temperature. They also may provide lift in the air while it is deftly maneuvering and hovering. These bats roost in caves and on cliffs and rock ledges. In the summer, the females form nursery colonies that may contain 1,000 bats.

| | |
|---|---|
| **SCIENTIFIC NAME** | *Corynorhinus townsendii* |
| **ANIMAL GROUP** | Mammals |
| **WINGSPAN/WEIGHT** | up to 13 in./0.5 oz. |
| **DIET** | carnivore: moths, flies, beetles |
| **LOCATION** | North America |
| **STATUS** | Least Concern |

• September 14th •
# WHITE-WINGED PETREL

Often seen feeding with other seabirds and dolphins, these birds take prey from or near to the surface of the sea. After breeding in rock crevices or tree cavities, the adults and young migrate together and spend the rest of the year at sea. Pairs return to the same nesting site every year to breed.

| | |
|---|---|
| **SCIENTIFIC NAME** | *Pterodroma leucoptera* |
| **ANIMAL GROUP** | Birds |
| **WINGSPAN/WEIGHT** | up to 2.5 ft./7 oz. |
| **DIET** | carnivore: squid, fish, krill |
| **LOCATION** | Pacific Ocean |
| **STATUS** | Vulnerable |

• September 15th •
# LEOPARD SHARK

Only found in the warm waters off the coasts of California and Mexico, this long, slim shark travels in schools along with rays and other sharks. It likes the sandy bottoms of bays and estuaries and uses the tidal waves to travel in and out. Its smooth, overlapping teeth crush the shells of its prey.

| | |
|---|---|
| **SCIENTIFIC NAME** | *Triakis semifasciata* |
| **ANIMAL GROUP** | Fish |
| **LENGTH/WEIGHT** | up to 6 ft./40 lb. |
| **DIET** | carnivore: crabs, shrimp, fish, worms |
| **LOCATION** | Pacific Ocean |
| **STATUS** | Least Concern |

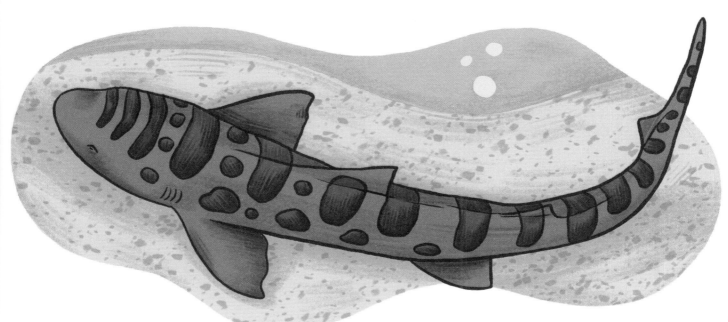

• September 16th •

# SNOWY OWL

Its wings spread wide, the owl swoops almost silently down to snatch up a fleeing lemming from the snowy ground. It had been perched nearby on its usual lookout point, listening for movement and turning its head to pinpoint exactly where the sounds were coming from. It can take prey from the ground, in the air or off the surface of water. It either swallows the prey whole or carries larger animals back to its perch or nest to tear into chunks.

| | |
|---|---|
| **SCIENTIFIC NAME** | *Bubo scandiacus* |
| **ANIMAL GROUP** | Birds |
| **WINGSPAN/WEIGHT** | up to 6 ft./7 lb. |
| **DIET** | carnivore: lemmings, voles, Arctic hares, ducks, fish |
| **LOCATION** | North America, northern Europe and northern Asia |
| **STATUS** | Vulnerable |

• September 17th •

# AFRICAN WILD DOG

This "painted dog" lives and hunts in a pack that contains up to 20 adults together with yearlings and pups. The adults use sight to find and hunt large prey, which they may chase for several miles at speeds of up to 35 mph.

| | |
|---|---|
| **SCIENTIFIC NAME** | Lycaon pictus |
| **ANIMAL GROUP** | Mammals |
| **LENGTH/WEIGHT** | up to 5 ft. including tail/80 lb. |
| **DIET** | carnivore: impala, gazelle, wildebeest, buffalo, rodents, birds |
| **LOCATION** | Africa |
| **STATUS** | Endangered |

• September 18th •

# BUDGETT'S FROG

Puffing up when threatened, this frog issues a loud scream from its enormous mouth to deter attackers. It bites with toothlike ridged jaws and two sharp fangs. It is a nocturnal sit-and-wait predator, submerging itself up to its nostrils in water until prey gets close.

| | |
|---|---|
| **SCIENTIFIC NAME** | Lepidobatrachus laevis |
| **ANIMAL GROUP** | Amphibians |
| **LENGTH/WEIGHT** | up to 5 in./8 oz. |
| **DIET** | carnivore: crickets, snails, worms, fish |
| **LOCATION** | central South America |
| **STATUS** | Least Concern |

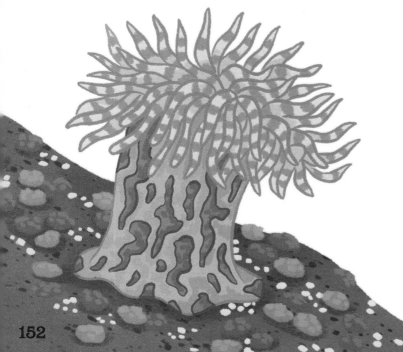

• September 19th •

# PAINTED ANEMONE

Sea anemones have stinging tentacles for defense, with crowns that can be more than 10 inches across. These invertebrates live under ledges and crevices in shallow waters, attached to large boulders or rock.

| | |
|---|---|
| **SCIENTIFIC NAME** | Urticina grebelnyi |
| **ANIMAL GROUP** | Invertebrates |
| **HEIGHT** | up to 20 in. |
| **DIET** | carnivore: shrimp, krill, mussels, fish |
| **LOCATION** | Atlantic and Pacific Oceans |
| **STATUS** | Not Evaluated |

• September 20th •

# SECRETARY BIRD

Walking more than it flies, this stately bird of prey makes its way across the African plains looking for snakes and lizards. It gets its name from the crest on its head that looks like an array of the quill pens that, when not in use, were often tucked behind the ears of 19th-century office workers. The secretary bird kills its prey by stamping on it – even a highly venomous puff adder or cobra – before eating it.

| | |
|---|---|
| **SCIENTIFIC NAME** | *Sagittarius serpentarius* |
| **ANIMAL GROUP** | Birds |
| **WINGSPAN/WEIGHT** | up to 7 ft./ 9 lb. |
| **DIET** | carnivore: snakes, insects, rats, lizards, birds |
| **LOCATION** | Africa |
| **STATUS** | Endangered |

# SURVIVING THE COLD

In the bitter cold, it is essential that animals stay warm if they are going to survive. There are different ways of dealing with the cold – whether it is a winter coat of thick fur or feathers, taking shelter or going into a deep sleep.

• September 21st •

## RAVEN

High metabolism (chemical reactions in the body) makes a lot of heat to keep this bird warm. So do special feathers on its nostrils and body feathers that fluff to trap warm air.

| | |
|---|---|
| SCIENTIFIC NAME | *Corvus corax* |
| ANIMAL GROUP | Birds |
| WINGSPAN/WEIGHT | up to 5 ft./4.4 lb. |
| DIET | omnivore: small animals, berries |
| LOCATION | Europe, Asia, North America, northern Africa |
| STATUS | Least Concern |

• September 22nd •

## ARCTIC HARE

To survive on the tundra, these hares have thick winter fur and paws that are heavily padded to insulate them from snow and ice. In the coldest weather, they huddle together in small snow dens.

| | |
|---|---|
| SCIENTIFIC NAME | *Lepus arcticus* |
| ANIMAL GROUP | Mammals |
| LENGTH/WEIGHT | up to 2.5 ft. including tail/15 lb. |
| DIET | omnivore: woody plants, mosses, berries, fish |
| LOCATION | northeastern North America |
| STATUS | Least Concern |

• September 23rd •

## GROUNDHOG

This sleeping beauty of the marmot family goes into hibernation (see pp.174–175) in its burrow for the whole winter. All of its body movements and other functions are massively reduced, and double layers of fur keep it warm and waterproof.

| | |
|---|---|
| SCIENTIFIC NAME | *Marmota monax* |
| ANIMAL GROUP | Mammals |
| LENGTH/WEIGHT | up to 2.6 ft. including tail/13 lb. |
| DIET | omnivore: grasses, leaves, tree bark, insects |
| LOCATION | North America |
| STATUS | Least Concern |

• September 24th •

# CARIBOU

These deer, also known as reindeer, have curled bones in their nostrils with blood vessels that warm icy air when it is breathed in. They can clear snow with their antlers, and their hooves have sharp footpads to cut into snow and ice when they walk, and into the earth when they dig for food.

**SCIENTIFIC NAME** | *Rangifer tarandus*
**ANIMAL GROUP** | Mammals
**LENGTH/WEIGHT** | up to 7 ft. including tail/660 lb.
**DIET** | herbivore: lichen, grasses, plants, fungi
**LOCATION** | northern North America, Europe and Asia
**STATUS** | Least Concern

• September 25th •

# BLACKFIN ICEFISH

This fish has antifreeze proteins in its ghostly white blood that stop ice from forming inside its body. It can survive in water temperatures of 28°F.

**SCIENTIFIC NAME** | *Chaenocephalus aceratus*
**ANIMAL GROUP** | Fish
**LENGTH/WEIGHT** | up to 2.3 ft./8 lb.
**DIET** | carnivore: fish, krill
**LOCATION** | Southern Ocean
**STATUS** | Not Evaluated

• September 26th •

# BELUGA WHALE

As it swims through Arctic waters, an insulating layer of blubber that is 100 times thicker than that of land mammals protects this whale. It has no dorsal fin because that might get damaged swimming among ice floes.

**SCIENTIFIC NAME** | *Delphinapterus leucas*
**ANIMAL GROUP** | Mammals
**LENGTH/WEIGHT** | up to 20 ft./1.8 tons
**DIET** | carnivore: octopus, squid, crabs, sandworms, fish
**LOCATION** | Arctic Ocean
**STATUS** | Least Concern

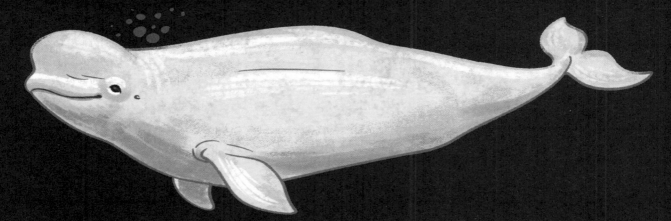

• September 27th •

# GOULDIAN FINCH

Also known as the rainbow finch, this is a wonderfully colored bird of meadows and plains close to water. It lives in eucalyptus trees and builds grass nests in hollows with both parent birds incubating the eggs and looking after the young. These finches are very social, and often flock with other birds.

| | |
|---|---|
| **SCIENTIFIC NAME** | *Chloebia gouldiae* |
| **ANIMAL GROUP** | Birds |
| **WINGSPAN/WEIGHT** | up to 6 in./0.5 oz. |
| **DIET** | omnivore: grass seeds, grains, insects |
| **LOCATION** | northern Australia |
| **STATUS** | Least Concern |

• September 28th •

# ACHALLO

This large-eared rodent, also known as the Altiplano chinchilla mouse, spends its day among boulders in rocky outcrops. The temperature is often freezing on the high plateaus of the mountains where it lives, so it has thick fur to keep it warm. It is an excellent climber, speeding up rocks and trees to escape hawks, eagles and snakes.

| | |
|---|---|
| **SCIENTIFIC NAME** | *Chinchillula sahamae* |
| **ANIMAL GROUP** | Mammals |
| **LENGTH/WEIGHT** | up to 11 in. including tail/4.7 oz. |
| **DIET** | herbivore: grasses, leaves, seeds |
| **LOCATION** | northwestern South America |
| **STATUS** | Least Concern |

• September 29th •

# EMPEROR SCORPION

One of the largest scorpions in the world, it lives in rainforests and savannas near the coast. There is a sting on the tip of its tail, but it mainly uses its palps (claws) to catch prey. Newborn scorpions, which are white and soft-bodied, travel on their mother's back for 10 to 20 days until their exoskeleton hardens.

| | |
|---|---|
| **SCIENTIFIC NAME** | *Pandinus imperator* |
| **ANIMAL GROUP** | Invertebrates |
| **LENGTH/WEIGHT** | up to 8 in./1.1 oz. |
| **DIET** | carnivore: insects, small rodents |
| **LOCATION** | western Africa |
| **STATUS** | Not Evaluated |

• September 30th •

# SPOTTED EAGLE RAY

These rays come into lagoons and estuaries, and swim around coral reefs searching in the mud for invertebrates. In open waters, they swim in large schools near the surface. If the ray is chased by a predator, it can leap out of the water, sometimes into a boat!

| | |
|---|---|
| **SCIENTIFIC NAME** | *Aetobatus narinari* |
| **ANIMAL GROUP** | Fish |
| **WIDTH/WEIGHT** | up to 11 ft./510 lb. |
| **DIET** | carnivore: clams, oysters, squid, fish |
| **LOCATION** | Atlantic, Pacific and Indian Oceans |
| **STATUS** | Endangered |

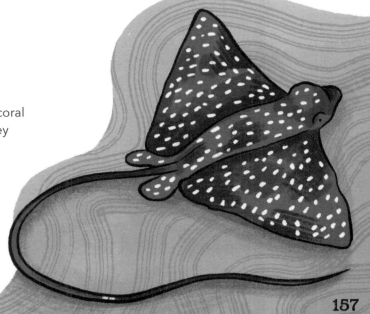

# OCTOBER

• October 1st •

## RED KANGAROO

These animals live in mobs - small family groups mainly of females and young, numbering around ten. The males "box" and kick each other when they are establishing who has mating rights. They are the largest marsupial, carrying their young in a pouch. When a baby is born, it is the size of a jellybean and climbs from the birth canal up its mother's fur to the pouch which it finally leaves eight months later.

| | |
|---|---|
| SCIENTIFIC NAME | *Osphranter rufus* |
| ANIMAL GROUP | Mammals |
| LENGTH/WEIGHT | up to 8 ft. including tail/200 lb. |
| DIET | herbivore: grasses, leaves, fruit, seeds |
| LOCATION | Australia |
| STATUS | Least Concern |

• October 2nd •

# GIANT TIGER LAND SNAIL

This is one of the largest land snails in the world. Its narrow shell is twice as long as it is wide and needs calcium for growth. So, although the snail is mainly a herbivore, it eats eggshells and bones when it finds them in the forest.

| | |
|---|---|
| **SCIENTIFIC NAME** | *Achatina achatina* |
| **ANIMAL GROUP** | Invertebrates |
| **LENGTH/WEIGHT** | up to 15 in./4.2 oz. |
| **DIET** | omnivore: leaves, flowers, stems, fruit, nuts, shells, bones |
| **LOCATION** | western Africa |
| **STATUS** | Not Evaluated |

• October 3rd •

# VICUÑA

The smallest member of the camel family, this animal needs to be fleet of foot because it is often hunted by pumas and condors (*see p.123*). It whistles to warn others when it is in danger. It lives high up in grasslands of the Andes Mountains and has a silky fleece that traps warm air and protects it from winter's freezing temperatures.

| | |
|---|---|
| **SCIENTIFIC NAME** | *Vicugna vicugna* |
| **ANIMAL GROUP** | Mammals |
| **LENGTH/WEIGHT** | up to 6 ft. including tail/145 lb. |
| **DIET** | herbivore: grasses, shrubs |
| **LOCATION** | South America |
| **STATUS** | Least Concern |

• October 4th •

# EUROPEAN FIRE-BELLIED TOAD

If attacked, this toad throws itself on its back to expose its "fire belly" to scare away the predator. If that does not work, it excretes poison through its skin that causes burning and sneezing. It lives in marshes or grassy wetlands and hibernates (*see pp.174-175*) under logs and among roots.

| | |
|---|---|
| **SCIENTIFIC NAME** | *Bombina bombina* |
| **ANIMAL GROUP** | Amphibians |
| **LENGTH/WEIGHT** | up to 2 in./0.5 oz. |
| **DIET** | carnivore: crickets, ants, worms |
| **LOCATION** | Europe |
| **STATUS** | Least Concern |

• October 5th •

# EMPEROR TAMARIN

Named after Wilhelm II of Germany who had a very splendid mustache, this small primate also has a tail much longer than its body. It lives in a family group of up to 15 in the middle canopy and treetops of the Amazon Rainforest, and will gouge tree trunks with its sharp claws to make gum flow to eat.

| | |
|---|---|
| **SCIENTIFIC NAME** | *Saguinus imperator* |
| **ANIMAL GROUP** | Mammals |
| **LENGTH/WEIGHT** | up to 2.2 ft. including tail/1.1 lb. |
| **DIET** | omnivore: gum, fruit, flowers, frogs, lizards, insects |
| **LOCATION** | northwestern South America |
| **STATUS** | Least Concern |

• October 6th •

## YELLOW-NAPED PARROT

With loud, screeching calls and whistles, large flocks of these excited birds fly over tropical forests. If stressed, they react by fanning their tails and rapidly contracting and expanding the pupils of their eyes, or eye pinning. These birds are good mimics and can learn to copy human words and even sentences.

| | |
|---|---|
| SCIENTIFIC NAME | *Amazona auropalliata* |
| ANIMAL GROUP | Birds |
| WINGSPAN/WEIGHT | up to 8 in./1.5 lb. |
| DIET | herbivore: seeds, nuts, fruit, leaves |
| LOCATION | Central America |
| STATUS | Critically Endangered |

• October 7th •

## SPINY LOBSTER

This clawless lobster lives in seagrass beds, and rocky and coral reefs in the Caribbean. It begins life in a nursery area on the shoreline, where it molts its old shell regularly and grows another one. When it matures, it heads for offshore reefs, migrating in a single-file line with up to 50 other lobsters.

| | |
|---|---|
| SCIENTIFIC NAME | *Panulirus argus* |
| ANIMAL GROUP | Invertebrates |
| LENGTH/WEIGHT | up to 18 in./10 lb. |
| DIET | carnivore: sea snails, mollusks, worms, shrimps |
| LOCATION | Atlantic Ocean |
| STATUS | Data Deficient |

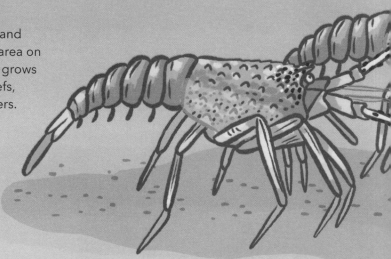

• October 8th •

# OKAPI

Deep in the Ituri Forest of the Democratic Republic of the Congo lives a solitary and well-camouflaged ungulate (hoofed mammal). It is the only living relative of the giraffe (see p.119). Its velvety fur is oily so rainwater slides off. Only the males have short horns and these slant back so they do not get tangled in vegetation.

| | |
|---|---|
| **SCIENTIFIC NAME** | *Okapia johnstoni* |
| **ANIMAL GROUP** | Mammals |
| **LENGTH/WEIGHT** | up to 9 ft. including tail/770 lb. |
| **DIET** | herbivore: leaves, twigs, fruit, clay |
| **LOCATION** | central Africa |
| **STATUS** | Endangered |

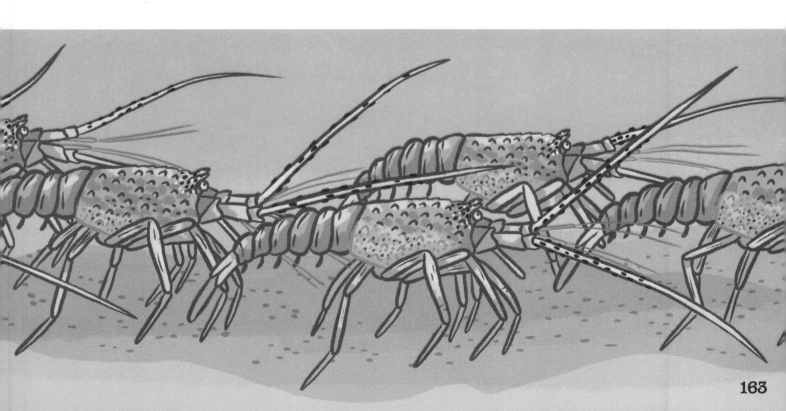

# NIGHTLIFE

Nocturnal animals are awake in the dark for a variety of reasons – perhaps escaping the heat of the day or avoiding predators. Many have developed amazing senses that help them to navigate in the dark while staying safe.

• October 9th •

## BIG DIPPER FIREFLY

Flitting through woodlands and over fields in the summer, the male dipper fireflies put on a display of long yellow flashes as they fly, to attract mates. Females respond with a single flash.

| | |
|---|---|
| SCIENTIFIC NAME | *Photinus pyralis* |
| ANIMAL GROUP | Invertebrates |
| LENGTH | up to 0.6 in. |
| DIET | omnivore: insects, nectar, slugs, worms |
| LOCATION | southeastern North America |
| STATUS | Least Concern |

• October 10th •

## TASMANIAN DEVIL

This marsupial's acute sense of smell helps it find prey at night. Its growls and blood-curdling screeches travel a long way on the night air, which is why it is called "devil."

| | |
|---|---|
| SCIENTIFIC NAME | *Sarcophilus harrisii* |
| ANIMAL GROUP | Mammals |
| LENGTH/WEIGHT | up to 3.6 ft. including tail/0.4 oz. |
| DIET | carnivore: wombats, wallabies, sheep, rabbits, carrion |
| LOCATION | Australia |
| STATUS | Endangered |

• October 11th •

## ANDEAN NIGHT MONKEY

Only found in the humid cloud forests of northern Peru, this monkey spends the night traveling and feeding in a small family group of up to six. It spends the day sleeping in tree hollows.

| | |
|---|---|
| SCIENTIFIC NAME | *Aotus miconax* |
| ANIMAL GROUP | Mammals |
| LENGTH/WEIGHT | up to 20 in. including tail/2.4 lb. |
| DIET | omnivore: fruit, flowers, leaves, insects |
| LOCATION | South America |
| STATUS | Endangered |

• October 12th •
# RONDO DWARF GALAGO

Also known as the Rondo bushbaby, this primate is only found in Tanzania. It has a tail that is longer than its body. Large eyes help it see in the dark and its big ears help it hear approaching danger.

| | |
|---|---|
| SCIENTIFIC NAME | *Paragalago rondoensis* |
| ANIMAL GROUP | Mammals |
| LENGTH/WEIGHT | up to 5 in. excluding tail/2.1 oz. |
| DIET | omnivore: insects, fruit, flowers |
| LOCATION | eastern Africa |
| STATUS | Endangered |

• October 13th •
# POLYPHEMUS MOTH

This large silk moth only eats when it is a caterpillar. It emerges from its cocoon in daylight so its wings strengthen in the sun, but then it becomes nocturnal and lives less than a week. The large spots in the middle of its hind wings imitate eyes, which puts off predators.

| | |
|---|---|
| SCIENTIFIC NAME | *Antheraea polyphemus* |
| ANIMAL GROUP | Invertebrates |
| WINGSPAN | up to 6 in. |
| DIET | herbivore: leaves, buds, shrubs |
| LOCATION | North America |
| STATUS | Least Concern |

• October 14th •
# NIGHTJAR

One of the best camouflaged of all the birds, the nightjar nests on the ground or perches motionless in trees during the day. At dusk and during the night, its large eyes easily spot insects on the wing that it captures with its wide mouth.

| | |
|---|---|
| SCIENTIFIC NAME | *Caprimulgus europaeus* |
| ANIMAL GROUP | Birds |
| WINGSPAN/WEIGHT | up to 2 ft./3.5 oz. |
| DIET | carnivore: flying insects, glowworms |
| LOCATION | Europe, Africa, Asia |
| STATUS | Least Concern |

• October 15th •

# OSTRICH

This flightless bird can run very fast, sprinting at up to 43 mph. With a neck that is almost half of its height, it is the world's largest and heaviest bird and lays the largest egg – equal in volume to two dozen hens' eggs! Ostriches are also excellent parents; males and females take turns all day to protect their young from predators such as hyenas (*see p.117*). They also use their bodies to cast shadows over the chicks in the intense heat of the sun and show them how to eat and graze for food as they move around.

| | |
|---|---|
| **SCIENTIFIC NAME** | *Struthio camelus* |
| **ANIMAL GROUP** | Birds |
| **HEIGHT/WEIGHT** | up to 9 ft./350 lb. |
| **DIET** | omnivore: roots, leaves, seeds, insects, lizards, rodents |
| **LOCATION** | Africa |
| **STATUS** | Least Concern |

• October 16th •

## GIANT OCEANIC MANTA RAY

This "devil fish" is the largest ray in the world. It is a filter-feeder but also hunts fish and breaches like the humpback whale (see p.74). It spends its time in the open ocean, diving deeply up to 3,300 feet to find its zooplankton prey. These rays have the largest brain in relation to body size of all known fish species.

| | |
|---|---|
| **SCIENTIFIC NAME** | *Mobula birostris* |
| **ANIMAL GROUP** | Fish |
| **WINGSPAN/WEIGHT** | up to 23 ft./2.2 tons |
| **DIET** | carnivore: zooplankton, fish, larvae |
| **LOCATION** | Atlantic, Pacific and Indian Oceans |
| **STATUS** | Endangered |

• October 17th •

# ILI PIKA

In the Tian Shan Mountains of northwestern China lives a cousin of the rabbit and hare. It makes its dens in rocky crevices at elevations from 9,500 to 13,000 feet, and does not hibernate. It collects hay from alpine meadows and makes piles for food that will get it through the cold of winter. It is a great climber, which it needs to be on the sloping bare rock of its craggy habitat.

| | |
|---|---|
| **SCIENTIFIC NAME** | *Ochotona iliensis* |
| **ANIMAL GROUP** | Mammals |
| **LENGTH/WEIGHT** | up to 8 in./9 oz. |
| **DIET** | herbivore: grasses, wildflowers, weeds |
| **LOCATION** | Asia |
| **STATUS** | Endangered |

• October 18th •

# SHOEBILL

If you were to imagine a bird, it is doubtful that it would be shoebill. Its head looks too large for its body, it has a shoe-shaped bill and amazingly wide wings given that it does not fly very often, and then not very far! This storklike bird lives in the swamps and marshlands of Africa, but does not have webbed feet. To keep cool, it gular flutters, opening its mouth and fluttering its neck muscles to lose heat.

| | |
|---|---|
| **SCIENTIFIC NAME** | *Balaeniceps rex* |
| **ANIMAL GROUP** | Birds |
| **WINGSPAN/WEIGHT** | up to 8 ft./15 lb. |
| **DIET** | carnivore: fish, water snakes, frogs, young crocodiles |
| **LOCATION** | central Africa |
| **STATUS** | Vulnerable |

• October 19th •

# DIAMONDBACK TERRAPIN

These terrapins are semiaquatic turtles that live in fresh or brackish (slightly salty) water and are named for the diamond pattern on their shell. They swim frequently but also spend time on land basking in the sun and burrowing in mud. The diamondback lives in the water of tidal marshes near the coast, swimming with its strongly webbed hind feet. It has glands near its eyes that flush out extra salt from its body.

**SCIENTIFIC NAME** | *Malaclemys terrapin*
**ANIMAL GROUP** | Reptiles
**LENGTH/WEIGHT** | up to 11 in./1.5 lb.
**DIET** | carnivore: snails, crustaceans, fish, insects
**LOCATION** | eastern North America
**STATUS** | Vulnerable

• October 20th •

# GOLDEN-SPOTTED TIGER BEETLE

This tiny, colorful beetle lives in sandy habitats, near rivers, on forest trails or in sand dunes near mangrove swamps. It is an efficient hunter with keen eyesight, and can quickly run down prey, even twice its size. Its larvae are ambush predators, lying in wait in vertical burrows.

**SCIENTIFIC NAME** | *Cicindela aurulenta*
**ANIMAL GROUP** | Invertebrates
**LENGTH** | up to 0.7 in.
**DIET** | carnivore: flies, beetles, caterpillars, spiders
**LOCATION** | Asia
**STATUS** | Not Evaluated

• October 21st •

# SUNDA COLUGO

While it is also known as the flying lemur, this animal is not a lemur, nor does it fly. However, it can glide up to 325 feet losing only 30 feet in height as it travels through dense tropical rainforest high up in the canopy. It does this using a furred flap of skin called the patagium that extends from its neck to its tail and the tips of its fingers and toes. Mother colugos wrap their young in the patagium like a blanket for protection and warmth.

| | |
|---|---|
| **SCIENTIFIC NAME** | *Galeopterus variegatus* |
| **ANIMAL GROUP** | Mammals |
| **LENGTH/WEIGHT** | up to 2.3 ft. including tail/4.4 lb. |
| **DIET** | omnivore: leaves, fruit, shoots, insects |
| **LOCATION** | southeastern Asia |
| **STATUS** | Least Concern |

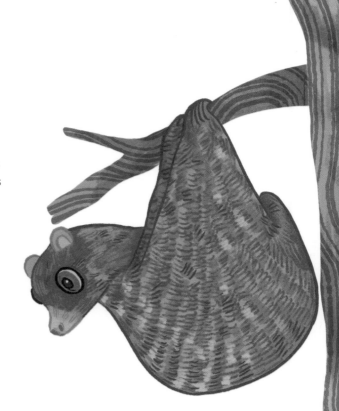

• October 22nd •

# GREAT BLUE-SPOTTED MUDSKIPPER

At low tide, this unusual air-breathing fish uses its fins to shuffle across mud flats, jumping with the aid of its tail if it wants to move more quickly. At high tide, it stays in a burrow it digs in the mud. Males are territorial, and confront rivals over females or burrows with open mouths.

| | |
|---|---|
| **SCIENTIFIC NAME** | *Boleophthalmus pectinirostris* |
| **ANIMAL GROUP** | Fish |
| **LENGTH/WEIGHT** | up to 9 in./3.2 oz. |
| **DIET** | herbivore: algae |
| **LOCATION** | Asia |
| **STATUS** | Not Evaluated |

- **October 23rd** -
# GRAY WOLF

Wolves howl to rally their pack, to hunt or to establish territory. Packs may number anything from 6 to 30, and are led by a breeding pair, the alpha male and female. They adapt easily to different habitats, from tundra to dense forest, and run down large prey over long distances with bursts of speed of up to 37 mph.

| | |
|---|---|
| **SCIENTIFIC NAME** | *Canis lupus* |
| **ANIMAL GROUP** | Mammals |
| **LENGTH/WEIGHT** | up to 7 ft. including tail/155 lb. |
| **DIET** | carnivore: elk, deer, wild boar |
| **LOCATION** | North America, Europe, Asia |
| **STATUS** | Least Concern |

• October 24th •

## BLACK WIDOW SPIDER

This is a solitary nocturnal hunter that flings silk over any struggling insect that gets caught in its web, wrapping it up. It then pulls the prey to one side of the web, injects it with poison and stores it to eat later. The red hourglass shape on its underside is a warning to predators, and it will drop out of the web and pretend it is dead if disturbed.

| | |
|---|---|
| **SCIENTIFIC NAME** | *Latrodectus mactans* |
| **ANIMAL GROUP** | Invertebrates |
| **LEGSPAN** | up to 2 in. |
| **DIET** | carnivore: insects, other arachnids |
| **LOCATION** | North America |
| **STATUS** | Not Evaluated |

• October 25th •

## NORTH ISLAND BROWN KIWI

Living on New Zealand's North Island, this nocturnal flightless bird lays the largest egg relative to its body weight. It has several burrows in its territory with tunnels up to 7 feet in length, each ending in a chamber in which the kiwi incubates the two eggs it lays at a time. When they hatch, the chicks are fully feathered and can leave the nest to fend for themselves only a week later.

| | |
|---|---|
| **SCIENTIFIC NAME** | *Apteryx mantelli* |
| **ANIMAL GROUP** | Birds |
| **LENGTH/WEIGHT** | up to 2.1 ft./9 lb. |
| **DIET** | omnivore: worms, beetles, snails, insects, berries |
| **LOCATION** | New Zealand |
| **STATUS** | Vulnerable |

# DORMANCY

Many animals need to slow down their bodily functions to become dormant or inactive because of environmental conditions. This may be for a few weeks, months or even years. There are different types of dormancy, including hibernation and estivation.

## • October 26th •
### RED-TAILED BUMBLEBEE

After the old queen has mated, she and the worker bees and males of a colony die. The new, young queens go into hibernation in the autumn and reappear the following spring to form colonies of their own.

| | |
|---|---|
| SCIENTIFIC NAME | *Bombus lapidarius* |
| ANIMAL GROUP | Invertebrates |
| LENGTH | up to 0.9 in. |
| DIET | herbivore: pollen, nectar |
| LOCATION | Europe |
| STATUS | Least Concern |

## • October 27th •
### EASTERN WATER-HOLDING FROG

This frog estivates when it becomes too hot and dry. It burrows underground during the dry season and makes itself a water-conserving cocoon from mucus. When it rains, maybe years later, it breaks out of it.

| | |
|---|---|
| SCIENTIFIC NAME | *Cyclorana platycephala* |
| ANIMAL GROUP | Amphibians |
| LENGTH | up to 2.8 in. |
| DIET | carnivore: insects, small fish |
| LOCATION | eastern Australia |
| STATUS | Least Concern |

## • October 28th •
### EASTERN BOX TURTLE

The turtle hibernates through the cold winters, usually for around six months. It makes a shallow burrow in woodland soil under leaf litter, its shell level with the soil's surface.

| | |
|---|---|
| SCIENTIFIC NAME | *Terrapene carolina* |
| ANIMAL GROUP | Reptiles |
| LENGTH/WEIGHT | up to 6 in./3.2 oz. |
| DIET | omnivore: fungi, berries, worms, slugs, insects |
| LOCATION | eastern North America |
| STATUS | Vulnerable |

## • October 29th •
# RED-SIDED GARTER SNAKE

When a cold-blooded animal hibernates it is called brumation. This snake does it for around six months when temperatures in Canada can be as low as minus 40°F. It brumates in very large numbers in natural burrows or under rocks.

| | |
|---|---|
| SCIENTIFIC NAME | Thamnophis sirtalis |
| ANIMAL GROUP | Reptiles |
| LENGTH/WEIGHT | up to 4 ft./2.6 oz. |
| DIET | carnivore: amphibians, worms, fish, small birds |
| LOCATION | North America |
| STATUS | Least Concern |

## • October 30th •
# FAT-TAILED DWARF LEMUR

On the island of Madagascar lives the world's only hibernating primate. It is not forced into hibernation because it is never very cold. However, between October and February, it is colder and there is less food around. So it stores fat in its tail and increases its body weight by up to 40 percent during the summer.

| | |
|---|---|
| SCIENTIFIC NAME | Cheirogaleus medius |
| ANIMAL GROUP | Mammals |
| LENGTH/WEIGHT | up to 20 in. including tail/10 oz. |
| DIET | omnivore: fruit, nectar, flowers, seeds, insects |
| LOCATION | Africa |
| STATUS | Vulnerable |

## • October 31st •
# ASIATIC BLACK BEAR

When it hibernates, this bear's heart rate drops to around 12 beats a minute (from 40–70) to conserve heat. It wakes easily, and like other bears, females give birth to cubs during this time.

| | |
|---|---|
| SCIENTIFIC NAME | Ursus thibetanus |
| ANIMAL GROUP | Mammals |
| LENGTH/WEIGHT | up to 7 ft./440 lb. |
| DIET | omnivore: honey, fruit, insects, small mammals |
| LOCATION | Asia |
| STATUS | Vulnerable |

# NOVEMBER

• November 1st •

## MARINE IGUANA

Looking like it dates back to the time of the dinosaurs, this extraordinary lizard is only found on, and in the seas around, the Galápagos Islands off the coast of Ecuador. Surprisingly for a creature that looks so fierce, it is a herbivore and, propelled by its flattened tail, searches for plants to eat both in the sea and on land.

| | |
|---|---|
| **SCIENTIFIC NAME** | *Amblyrhynchus cristatus* |
| **ANIMAL GROUP** | Reptiles |
| **LENGTH/WEIGHT** | up to 5 ft. including tail/3.3 lb. |
| **DIET** | herbivore: algae, seaweed, coastal plants |
| **LOCATION** | South America |
| **STATUS** | Vulnerable |

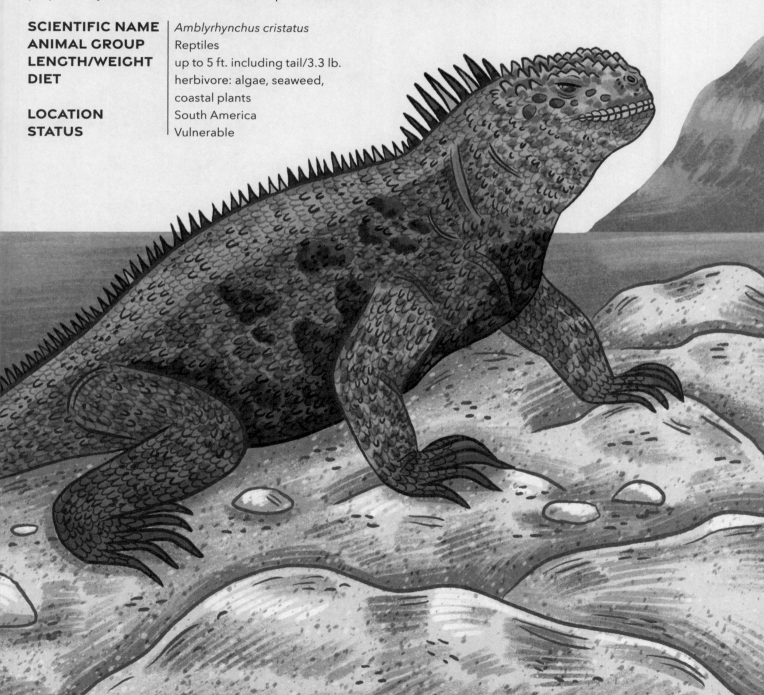

• November 2nd •

# WILD BOAR

A thickset forest animal with a very large head that takes up to one-third of its length, this wild pig uses its long snout to forage in the ground for worms and bugs. Both boars (males) and sows (females) have tusks that they use for defense. The piglets look very different from their parents. They are ginger-brown and have striped coats that help camouflage them on the forest floor.

| | |
|---|---|
| **SCIENTIFIC NAME** | *Sus scrofa* |
| **ANIMAL GROUP** | Mammals |
| **LENGTH/WEIGHT** | up to 8 ft. including tail/660 lb. |
| **DIET** | omnivore: fruit, roots, leaves, insects, birds, small mammals |
| **LOCATION** | Europe, Asia, northwestern Africa |
| **STATUS** | Least Concern |

• November 3rd •

# GREAT HORNBILL

This noisy bird has a very large bill, and a hornlike growth called a casque on the top of its head. During the day, it hops along the branches of rainforest trees searching for fruit and insects. Anything it catches, it tosses in the air and swallows whole. At night, large numbers of hornbills roost together on the highest trees.

| | |
|---|---|
| **SCIENTIFIC NAME** | *Buceros bicornis* |
| **ANIMAL GROUP** | Birds |
| **WINGSPAN/WEIGHT** | up to 6 ft./9 lb. |
| **DIET** | omnivore: mainly fruit, also small animals, insects |
| **LOCATION** | southern and southeastern Asia |
| **STATUS** | Vulnerable |

• November 4th •

# LONGSPINE PORCUPINE FISH

Living on shallow reefs and in seagrass beds near the sea floor, this fish's sharp beak crushes the shells of the crustaceans it hunts at night. It has a unique way of protecting itself from predators. If it is threatened, it takes in water and inflates itself to two or three times its normal size. This makes the spikes on its body stand out.

**SCIENTIFIC NAME** *Diodon holocanthus*
**ANIMAL GROUP** Fish
**LENGTH** up to 2 ft.
**DIET** carnivore: hermit crabs, limpets, sea snails
**LOCATION** tropical oceans worldwide
**STATUS** Least Concern

• November 5th •

# GREEN ANACONDA

As an ambush predator, there is nothing this heavy snake likes more than a capybara (see p.97) that comes down to the water to drink in the early evening. The snake's eyes are on the top of its head so they can keep watch while the rest of it is hidden in the water. It is a constrictor, so after catching the prey in its jaws, it winds around the animal, then tightens its muscles to crush it.

**SCIENTIFIC NAME** *Eunectes murinus*
**ANIMAL GROUP** Reptiles
**LENGTH/WEIGHT** up to 30 ft./550 lb.
**DIET** carnivore: fish, reptiles, birds, capybaras
**LOCATION** South America
**STATUS** Least Concern

# URBAN ANIMALS

More and more wild animals are moving into towns and cities all over the world as their habitats are taken over or destroyed, and they have adapted to living with people, traffic and buildings.

• November 6th •

## MANDARIN DUCK

Now a common sight in parks and green urban areas all over the world, this colorful, migratory bird's natural habitat in the wild is forests with fast-flowing rocky streams.

| | |
|---|---|
| SCIENTIFIC NAME | *Aix galericulata* |
| ANIMAL GROUP | Birds |
| WINGSPAN/WEIGHT | up to 30 in./1.5 lb. |
| DIET | omnivore: plants, seeds, insects |
| LOCATION | eastern Asia |
| STATUS | Least Concern |

• November 7th •

## TUFTED GRAY LANGUR

This monkey has moved into built-up areas of cities and towns in India and Sri Lanka. Tall buildings pose no difficulty to a primate that is used to living in the tall trees of tropical rainforests.

| | |
|---|---|
| SCIENTIFIC NAME | *Semnopithecus priam* |
| ANIMAL GROUP | Mammals |
| LENGTH/WEIGHT | up to 6 ft. including tail/40 lb. |
| DIET | herbivore: leaves, shoots, grass, bamboo |
| LOCATION | Asia |
| STATUS | Near Threatened |

• November 8th •

## SIKA DEER

In the Japanese city of Nara, more than 1,200 wild sika deer roam the streets freely from dusk to dawn. They live in the park, but wander wherever they want – through temples, shops and subway stations.

| | |
|---|---|
| SCIENTIFIC NAME | *Cervus nippon* |
| ANIMAL GROUP | Mammals |
| LENGTH/WEIGHT | up to 7 ft. including tail/155 lb. |
| DIET | herbivore: grasses, heather, fungi |
| LOCATION | eastern Asia |
| STATUS | Least Concern |

• November 9th •
# PEREGRINE FALCON

This bird can be found on the tops of high buildings in cities all over the world. Its natural nest site is a cliff ledge, but skyscrapers provide the same amount of privacy and access to the bird's main prey, the pigeon.

| | |
|---|---|
| **SCIENTIFIC NAME** | *Falco peregrinus* |
| **ANIMAL GROUP** | Birds |
| **WINGSPAN/WEIGHT** | up to 4 ft./3.3 lb. |
| **DIET** | carnivore: birds, rabbits, bats |
| **LOCATION** | worldwide except Antarctica |
| **STATUS** | Least Concern |

• November 10th •
# ROCK DOVE

Also known as the feral pigeon in an urban setting, this bird lives in towns and cities everywhere in the world. It gathers in flocks of many hundreds of birds, and often nests on roof ledges or inside the attics of houses.

| | |
|---|---|
| **SCIENTIFIC NAME** | *Columba livia* |
| **ANIMAL GROUP** | Birds |
| **WINGSPAN/WEIGHT** | up to 2.3 ft./13 oz. |
| **DIET** | omnivore: grains, leaves, insects |
| **LOCATION** | Europe, North Africa, southwestern Asia |
| **STATUS** | Least Concern |

• November 11th •
# RED FOX

No animal has settled more successfully into a suburban way of life than the red fox. It is thought to be the most common non-domestic carnivore in cities. It finds food easily and digs its dens under sheds in gardens.

| | |
|---|---|
| **SCIENTIFIC NAME** | *Vulpes vulpes* |
| **ANIMAL GROUP** | Mammals |
| **LENGTH/WEIGHT** | up to 4.6 ft. including tail/35 lb. |
| **DIET** | omnivore: small animals, insects, worms, fruit |
| **LOCATION** | North America, Europe, Asia, northern Africa, Australia |
| **STATUS** | Least Concern |

• November 12th •

# STRIPED SKUNK

When a skunk raises its tail, beware! If it cannot run away from danger, and stamping on the ground as a warning does not work, its final defense is to spray a strong-smelling yellow liquid from stink glands on its back end. The spray can travel up to 20 feet and be smelled over half a mile away. It will burn eyes and give the skunk time to escape. This solitary, nocturnal animal is immune to snake venom.

| | |
|---|---|
| **SCIENTIFIC NAME** | *Mephitis mephitis* |
| **ANIMAL GROUP** | Mammals |
| **LENGTH/WEIGHT** | up to 35 in. including tail/14 lb. |
| **DIET** | omnivore: grasshoppers, beetles, birds, snakes, plants |
| **LOCATION** | North America |
| **STATUS** | Least Concern |

• November 13th •

# ATLANTIC PUFFIN

Diving as deep as 130 feet into the sea, this puffin can catch several small fish, holding them in its beak, which is serrated to keep the sand eels or sprats in place while the bird gathers more on the same dive. It carries the fish back to the shore to eat itself or give to its puffling – it has just one chick a year with the same mate.

| | |
|---|---|
| **SCIENTIFIC NAME** | *Fratercula arctica* |
| **ANIMAL GROUP** | Birds |
| **WINGSPAN/WEIGHT** | up to 2 ft./1 lb. |
| **DIET** | carnivore: fish, crustaceans, mollusks |
| **LOCATION** | northeastern North America, northern and western Europe |
| **STATUS** | Vulnerable |

• November 14th •

# LEAF-CUTTER ANT

These busy insects live on the floor of the rainforest, climbing into the canopy to collect leaves that they cut with their sharp jaws. They carry pieces up to 50 times their own weight, in long lines back to their underground nest, where they use the decaying leaves to grow fungi to eat. Each nest may contain millions of ants and stretch over more than 33 square feet.

| | |
|---|---|
| **SCIENTIFIC NAME** | *Atta cephalotes* |
| **ANIMAL GROUP** | Invertebrates |
| **LENGTH** | up to 1 in. |
| **DIET** | herbivore: fungi |
| **LOCATION** | Central and South America |
| **STATUS** | Not Evaluated |

• November 15th •

# INDIAN BULLFROG

Also known as the golden frog, the males turn a vibrant yellow color with bright blue vocal sacs during the breeding season. At other times they are olive green or brown, with a yellow streak along their spines. They live in wetlands, emerging from their hidden burrows when the South Asian monsoon rains arrive. Having changed into mating colors, the males sing out loudly to attract females.

| | |
|---|---|
| **SCIENTIFIC NAME** | *Hoplobatrachus tigerinus* |
| **ANIMAL GROUP** | Amphibians |
| **LENGTH/WEIGHT** | up to 7 in./1.3 lb. |
| **DIET** | carnivore: mice, insects, worms, snakes, birds |
| **LOCATION** | southern Asia |
| **STATUS** | Least Concern |

• November 16th •

# RED DEER

Every year during the rutting season, a period of two months in the autumn, red deer stags engage in dramatic contests for hinds. The stags bellow their mating calls, stake out their territory, and clash their huge antlers with rival males. These deer live in small herds of hinds and their young. The stags are usually solitary but gather females in the rutting season.

| | |
|---|---|
| **SCIENTIFIC NAME** | *Cervus elaphus* |
| **ANIMAL GROUP** | Mammals |
| **LENGTH/WEIGHT** | up to 9 ft. including tail/750 lb. |
| **DIET** | herbivore: tree shoots, grasses, fruit, seeds |
| **LOCATION** | Europe |
| **STATUS** | Least Concern |

# OUT OF SIGHT

Burrowing is a technique used by many different animals. They may tunnel into the hardest of surfaces to make themselves dens to sleep and nest in, for protection from predators, to store food or to set a trap for prey.

• November 17th •

## COMMON WOMBAT

Rarely seen, this marsupial spends up to 16 hours a day sleeping in its burrow, emerging at night to graze when it is cool.

| | |
|---|---|
| SCIENTIFIC NAME | *Vombatus ursinus* |
| ANIMAL GROUP | Mammals |
| LENGTH/WEIGHT | up to 3.9 ft./80 lb. |
| DIET | herbivore: grasses, leaves, moss, roots, tubers, bark |
| LOCATION | southeastern Australia |
| STATUS | Least Concern |

• November 18th •

## BURROWING OWL

This burrowing bird is the only small owl that perches on the ground. It hunts during the day, running in search of prey. It digs its own burrows but will also take over abandoned dens.

| | |
|---|---|
| SCIENTIFIC NAME | *Athene cunicularia* |
| ANIMAL GROUP | Birds |
| WINGSPAN/WEIGHT | up to 2 ft./9 oz. |
| DIET | carnivore: insects, small rodents, birds, lizards |
| LOCATION | North and South America |
| STATUS | Least Concern |

• November 19th •

## SEA URCHIN

Known as the burrowing urchin, it is found on the sea floor in tropical ocean reefs. It scrapes its way through solid rock and coral with its teeth and spines to make itself a safe place to hide.

| | |
|---|---|
| SCIENTIFIC NAME | *Echinometra mathaei* |
| ANIMAL GROUP | Invertebrates |
| WIDTH | up to 3 in. |
| DIET | omnivore: algae, small invertebrates |
| LOCATION | Indian and Pacific Oceans |
| STATUS | Not Evaluated |

• November 20th •
# FUNNEL-WEB SPIDER

In damp forest regions, this spider burrows into moist, sheltered places in rotting logs or under rocks. It sets up silk trip-lines radiating out from its burrow entrance. It places its front legs on the trip-lines to detect prey, which it subdues with venom.

| | |
|---|---|
| SCIENTIFIC NAME | *Atrax robustus* |
| ANIMAL GROUP | Invertebrates |
| LENGTH | up to 1.4 in. |
| DIET | carnivore: beetles, cockroaches, skinks, snails |
| LOCATION | eastern Australia |
| STATUS | Not Evaluated |

• November 21st •
# DESERT POCKET GOPHER

Living in a burrow that has passages and chambers, this rodent leaves sandy mounds above ground as it digs, like a mole. When it returns from foraging expeditions, it plugs the burrow opening to control the temperature inside and keep out predators.

| | |
|---|---|
| SCIENTIFIC NAME | *Geomys arenarius* |
| ANIMAL GROUP | Mammals |
| LENGTH/WEIGHT | up to 12 in. including tail/9 oz. |
| DIET | herbivore: roots, tubers, seeds, fruit |
| LOCATION | North America |
| STATUS | Near Threatened |

• November 22nd •
# LEAST WEASEL

This predator is not very good at digging, but instead nests in the abandoned burrow of another animal, often that of its prey. It will dig a small area near its den to store leftover food from its kills, so that when food is scarce it has plenty to eat.

| | |
|---|---|
| SCIENTIFIC NAME | *Mustela nivalis* |
| ANIMAL GROUP | Mammals |
| LENGTH/WEIGHT | up to 14 in. including tail/7 oz. |
| DIET | carnivore: mice, voles, young rabbits, birds, eggs |
| LOCATION | North America, Europe, Asia |
| STATUS | Least Concern |

• November 23rd •

# GIANT ANTEATER

Also known as the ant bear, this enormous mammal walks slowly across a grassy plain toward a large termite mound. It visits up to 200 termite or ant nests a day in a quest to find the 30,000-odd insects it needs to eat daily. It uses its claws to open the nests, sticks its long snout inside and slurps up the insects with its 20-inch-long sticky tongue.

| | |
|---|---|
| **SCIENTIFIC NAME** | *Myrmecophaga tridactyla* |
| **ANIMAL GROUP** | Mammals |
| **LENGTH/WEIGHT** | up to 8 ft. including tail/120 lb. |
| **DIET** | carnivore: ants, termites |
| **LOCATION** | Central and South America |
| **STATUS** | Vulnerable |

• November 24th •

# GOLIATH TARANTULA

Deep in the rainforest, in a silk-lined burrow and under rocks and roots, lives a solitary spider that is the largest in the world, often as big as a dinner plate. It is mostly active at night. As a defense, it rubs its hind legs on its abdomen to fire long bristles at a predator. These can embed themselves and irritate, itch and burn, particularly if they hit the eyes.

| | |
|---|---|
| **SCIENTIFIC NAME** | *Theraphosa blondi* |
| **ANIMAL GROUP** | Invertebrates |
| **LEGSPAN/WEIGHT** | up to 11 in./6 oz. |
| **DIET** | carnivore: rodents, frogs, lizards, snakes |
| **LOCATION** | northern South America |
| **STATUS** | Not Evaluated |

• November 25th •

# KOMODO DRAGON

This monitor lizard only lives on the Sunda Islands of Indonesia. It is a fierce hunter, often in a group, and the apex predator of the islands, with a venomous bite that stops its prey's blood from clotting. It will even eat hatchlings of its own species, so it is no surprise that young Komodo dragons spend most of their time up trees, where they cannot be caught and eaten by an adult.

| | |
|---|---|
| SCIENTIFIC NAME | *Varanus komodoensis* |
| ANIMAL GROUP | Reptiles |
| LENGTH/WEIGHT | up to 10 ft. including tail/365 lb. |
| DIET | carnivore: pigs, deer, carrion |
| LOCATION | southeastern Asia |
| STATUS | Endangered |

• November 26th •

# LARGETOOTH SAWFISH

This ray lives in shallow waters in estuaries, bays and around the coast. Its saw has teeth all the way up each edge and is used to stir up the seabed looking for crustaceans to eat. It can also treat the saw as a weapon to stun small fish. It often lies on the sandy seabed, breathing by drawing water into its gills through spiracles – large holes behind each eye.

| | |
|---|---|
| SCIENTIFIC NAME | *Pristis pristis* |
| ANIMAL GROUP | Fish |
| LENGTH/WEIGHT | up to 21 ft./1,325 lb. |
| DIET | carnivore: crustaceans, small fish, mollusks |
| LOCATION | worldwide in warm water oceans |
| STATUS | Critically Endangered |

• November 27th •

# GREATER ROADRUNNER

A familiar sight running along roads in the southern USA, this is a long-legged member of the cuckoo family. With two toes pointing forward and two pointing backward, it can run at up to 19 mph. It only goes airborne if it is threatened and even then it needs a running start. It forages in deserts and grasslands for food, and can kill and swallow a rattlesnake whole.

| | |
|---|---|
| SCIENTIFIC NAME | *Geococcyx californianus* |
| ANIMAL GROUP | Birds |
| WINGSPAN/WEIGHT | up to 2 ft./1.2 lb. |
| DIET | omnivore: lizards, hummingbirds, insects, scorpions, spiders, snakes, prickly pear cactus, fruit |
| LOCATION | southern North America |
| STATUS | Least Concern |

• November 28th •

# WESTERN HONEYBEE

The most common of the honeybees, its Latin name "mellifera" means "honey-bearing." It is a vital pollinator of fruit, flowers and vegetables. When the worker bees collect nectar from flowers, yellow pollen dust rubs off on their bodies and is carried to the next flower to pollinate it.

| | |
|---|---|
| **SCIENTIFIC NAME** | Apis mellifera |
| **ANIMAL GROUP** | Invertebrates |
| **LENGTH** | up to 0.8 in. |
| **DIET** | herbivore: pollen, nectar, honey |
| **LOCATION** | Europe, Africa, western Asia |
| **STATUS** | Data Deficient |

• November 29th •

# FRIED EGG JELLYFISH

These extraordinary and aptly named creatures can be seen in summer when the waters of the seas they live in are at their warmest. Huge "blooms" of thousands appear near the surface.

| | |
|---|---|
| **SCIENTIFIC NAME** | Cotylorhiza tuberculata |
| **ANIMAL GROUP** | Invertebrates |
| **WIDTH** | up to 16 in. |
| **DIET** | omnivore: zooplankton, phytoplankton |
| **LOCATION** | western Atlantic Ocean, Mediterranean Sea |
| **STATUS** | Not Evaluated |

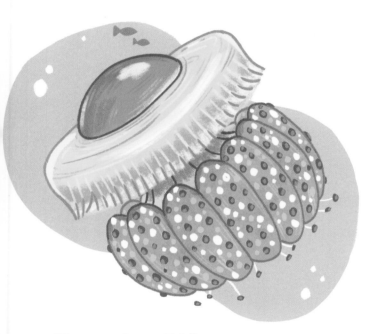

• November 30th •

# MACLEAY'S SPECTRE

Also known as the Australian walking stick, this giant stick insect has very effective camouflage, looking just like dried leaves. If it is threatened, it lifts its front legs and curls up its abdomen in what is described as a "scorpion" pose.

| | |
|---|---|
| **SCIENTIFIC NAME** | Extatosoma tiaratum |
| **ANIMAL GROUP** | Invertebrates |
| **LENGTH/WEIGHT** | up to 6 in./0.9 oz. |
| **DIET** | herbivore: leaves of rose, eucalyptus, oak, raspberry, blackberry, hazel |
| **LOCATION** | eastern Australia |
| **STATUS** | Least Concern |

# DECEMBER

• December 1st •

## BEARDED VULTURE

Also known as the lammergeier, this is the second largest of the vultures with a massive wingspan. This scavenger is a bone-eater, feeding almost wholly on carrion, and the only bird that can digest bones. Swallowing smaller bones whole, it drops the larger ones from a height onto the rocks to break them. When it does take live prey, such as tortoises and lambs, it uses the same method to kill them.

| | |
|---|---|
| SCIENTIFIC NAME | *Gypaetus barbatus* |
| ANIMAL GROUP | Birds |
| WINGSPAN/WEIGHT | up to 9 ft./15 lb. |
| DIET | carnivore: bones, carrion, birds, mammals, tortoises, reptiles |
| LOCATION | Europe, Asia, eastern Africa |
| STATUS | Near Threatened |

### • December 2nd •
# BHUTAN GLORY

This beautiful insect is a swallowtail butterfly that lives in the forested hills and mountains of Bhutan and other parts of Asia. Its caterpillars eat leaves while they are growing. After emerging, the butterfly lives for around 180 days, an extremely long time for a butterfly, flying from flower to flower and feeding on nectar.

| | |
|---|---|
| **SCIENTIFIC NAME** | *Bhutanitis lidderdalii* |
| **ANIMAL GROUP** | Invertebrates |
| **WINGSPAN** | up to 4.3 in. |
| **DIET** | herbivore: leaves, stalks, nectar |
| **LOCATION** | southeastern Asia |
| **STATUS** | Least Concern |

### • December 3rd •
# TOCO TOUCAN

The largest of all the toucans, the toco also has the biggest beak relative to its body size – one-third of its length. The beak is lightweight because it is hollow and made of keratin (like your fingernails). The toco mainly eats fruit such as figs, oranges and guava, peeling them with its beak, tipping its head back and swallowing.

| | |
|---|---|
| **SCIENTIFIC NAME** | *Ramphastos toco* |
| **ANIMAL GROUP** | Birds |
| **WINGSPAN/WEIGHT** | up to 4.9 ft./1.9 lb. |
| **DIET** | omnivore: mainly fruit, also insects, eggs |
| **LOCATION** | South America |
| **STATUS** | Least Concern |

• December 4th •

# PLAINS ZEBRA

Every zebra has unique stripes, so it is easy to identify. They live in small family groups with one stallion, several mares and their foals on grasslands and in savanna woodlands. Groups groom and play with one another, communicating with calls and facial expressions.

| | |
|---|---|
| **SCIENTIFIC NAME** | *Equus quagga* |
| **ANIMAL GROUP** | Mammals |
| **LENGTH/WEIGHT** | up to 10 ft. including tail/850 lb. |
| **DIET** | herbivore: mainly grass, also leaves, bark, roots |
| **LOCATION** | southern and eastern Africa |
| **STATUS** | Near Threatened |

• December 5th •

# RED PANDA

Like the giant panda (see p.95), this is a bamboo-eater. It lives in the high-altitude forests of the eastern Himalayas. However, it is not a bear and is also not related to the panda. It is the only member of its family and is solitary apart from the mating season. Its long, bushy tail helps it balance in trees and wraps round its body for warmth. In very cold weather, it becomes dormant (see pp.174–175).

| | |
|---|---|
| **SCIENTIFIC NAME** | *Ailurus fulgens* |
| **ANIMAL GROUP** | Mammals |
| **LENGTH/WEIGHT** | up to 3.6 ft. including tail/17 lb. |
| **DIET** | omnivore: bamboo, grasses, fruit, grubs, birds, small mammals |
| **LOCATION** | Asia |
| **STATUS** | Endangered |

# ANIMAL SUPERPOWERS

Whether they are swifter than most, defy gravity or live forever, some animals take your breath away with their superpowers! Read on to find out about some truly mind-blowing animals with extraordinary abilities.

• December 6th •

## PISTOL SHRIMP

Snapping the larger of its two claws, this tiny shrimp fires a bubble that is nearly as hot as the surface of the sun and makes a sound louder than a gunshot, stunning prey and predators alike.

| | |
|---|---|
| SCIENTIFIC NAME | *Alpheus randalli* |
| ANIMAL GROUP | Invertebrates |
| LENGTH | up to 2 in. |
| DIET | carnivore: fish, small invertebrates |
| LOCATION | Indian and Pacific Oceans |
| STATUS | Not Evaluated |

• December 7th •

## SHORTFIN MAKO

Streamlined for speed, this is the world's fastest shark with speeds of up to 37 mph. It is unstoppable, undulating dynamically through the open ocean.

| | |
|---|---|
| SCIENTIFIC NAME | *Isurus oxyrinchus* |
| ANIMAL GROUP | Fish |
| LENGTH/WEIGHT | up to 13 ft./1,100 lb. |
| DIET | carnivore: tuna, squid, swordfish, smaller sharks |
| LOCATION | warm oceans worldwide |
| STATUS | Endangered |

• December 8th •

## COMMON OPOSSUM

This prey animal, which is not very fast, has no real defenses, but it has a surprising ability to kill and eat poisonous snakes. Its blood is known to neutralize the venom of some snakes, such as the western back rattlesnake.

| | |
|---|---|
| SCIENTIFIC NAME | *Didelphis virginiana* |
| ANIMAL GROUP | Mammals |
| LENGTH/WEIGHT | up to 3.6 ft. including tail/14 lb. |
| DIET | omnivore: fruit, insects, frogs, small mammals |
| LOCATION | North America, Central America |
| STATUS | Least Concern |

• December 9th •
# ORCA

Pods (groups) of these intelligent hunters herd fish, stunning the prey by slapping their tails on the surface of the water. They also ram ice floes to jerk seals into the water.

| | |
|---|---|
| **SCIENTIFIC NAME** | *Orcinus orca* |
| **ANIMAL GROUP** | Mammals |
| **LENGTH/WEIGHT** | up to 30 ft./11 tons |
| **DIET** | carnivore: marine mammals, fish, seabirds, turtles |
| **LOCATION** | oceans worldwide |
| **STATUS** | Data Deficient |

• December 10th •
# ALPINE IBEX

Able to climb the most sheer and terrifying rock surfaces to reach mineral licks, this wild goat defies gravity. Ibex kids are natural rock climbers too and follow their mothers up the steep rocks from when they are only one day old!

| | |
|---|---|
| **SCIENTIFIC NAME** | *Capra ibex* |
| **ANIMAL GROUP** | Mammals |
| **LENGTH/WEIGHT** | up to 6 ft. including tail/265 lb. |
| **DIET** | herbivore: grasses, flowers, twigs, leaves |
| **LOCATION** | Europe |
| **STATUS** | Least Concern |

• December 11th •
# IMMORTAL JELLYFISH

When injured or stressed, the adult goes down to the ocean floor and morphs back into infancy – a polyp. In two months, it is a new jellyfish, and it can do this over and over again.

| | |
|---|---|
| **SCIENTIFIC NAME** | *Turritopsis dohrnii* |
| **ANIMAL GROUP** | Invertebrates |
| **WIDTH** | up to 0.2 in. |
| **DIET** | carnivore: zooplankton, fish eggs, small mollusks |
| **LOCATION** | warm oceans worldwide |
| **STATUS** | Not Evaluated |

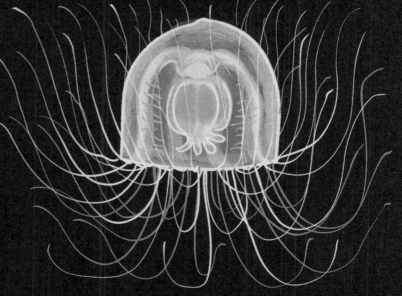

• December 12th •

# COMMON MARMOSET

Also called the white-tufted marmoset, this small primate has a tail longer than its body. It inhabits trees at the edges of the rainforest and plantations in Brazil, where it lives in a small family group that includes a breeding pair. The care of their young as they grow is shared with the other relations in the group.

| | |
|---|---|
| **SCIENTIFIC NAME** | *Callithrix jacchus* |
| **ANIMAL GROUP** | Mammals |
| **LENGTH/WEIGHT** | up to 20 in. including tail/9 oz. |
| **DIET** | omnivore: tree sap, insects, fruit, nectar, lizards, frogs |
| **LOCATION** | northeastern South America |
| **STATUS** | Least Concern |

• December 13th •

# COMMON CUTTLEFISH

With a light show in which it flashes several colors rapidly and undulates its body, the cuttlefish distracts a fish. It shoots out its two feeding tentacles to capture the prey, then grasps it with the suckers on its eight arms to guide it to its parrotlike beak. Cuttlefish, like squid and octopuses, have three hearts. They expel ink to cloud the water, and can also move away quickly using jet propulsion, forcing water out of their bodies to shoot forward.

| | |
|---|---|
| **SCIENTIFIC NAME** | *Sepia officinalis* |
| **ANIMAL GROUP** | Invertebrates |
| **LENGTH/WEIGHT** | up to 18 in./9 lb. |
| **DIET** | carnivore: fish, mollusks, crabs, shrimps |
| **LOCATION** | northeastern Atlantic Ocean, European seas |
| **STATUS** | Least Concern |

• December 14th •

# AMERICAN BISON

These powerful animals, also known as buffalo, are the largest and heaviest animals in North America. They are surprisingly quick on their feet, reaching speeds of 34 mph. They are agile, turn quickly, jump over obstacles and swim strongly. In prehistoric times, they roamed over the grasslands and prairies in their millions, but today there are only around 31,000, mainly in national parks.

| | |
|---|---|
| **SCIENTIFIC NAME** | *Bison bison* |
| **ANIMAL GROUP** | Mammals |
| **LENGTH/WEIGHT** | up to 12 ft. including tail/ 1.1 tons |
| **DIET** | herbivore: grasses, lichen, flowering plants, leaves |
| **LOCATION** | North America |
| **STATUS** | Near Threatened |

• December 15th •

# HARBOR PORPOISE

Staying close to coastal areas or river estuaries, this is one of the smallest cetaceans. It often travels up rivers as well, and has been found hundreds of miles from the sea. It is normally solitary or travels in small groups, but up to 50 may gather to feed or migrate (see pp.84-85). When they surface and breathe out, the sound is like a sneeze, so fishers sometimes call them "puffers."

| | |
|---|---|
| **SCIENTIFIC NAME** | *Phocoena phocoena* |
| **ANIMAL GROUP** | Mammals |
| **LENGTH/WEIGHT** | up to 7 ft./170 lb. |
| **DIET** | carnivore: fish, sand eels, squid, octopuses |
| **LOCATION** | northern Pacific and Atlantic Oceans, inland seas of Europe and Asia |
| **STATUS** | Least Concern |

• December 16th •

# WESTERN LOWLAND GORILLA

These social great apes live in troops that number up to ten, and are composed of a large silverback male, several adult females and their young. They forage in the day and build new nests to sleep in every night. Gorillas display many humanlike emotions, including laughter to show pleasure. The young learn from the adults but also play together, usually wrestling and biting. Although these gorillas can climb trees, they spend most of their time on the ground, and when they move around the lowland forests, swamps and marshes where they live, they knuckle-walk, supporting their weight on their curled hands.

| | |
|---|---|
| **SCIENTIFIC NAME** | *Gorilla gorilla* |
| **ANIMAL GROUP** | Mammals |
| **HEIGHT/WEIGHT** | up to 6 ft./500 lb. |
| **DIET** | omnivore: leaves, stems, vines, fruit, ants, termites |
| **LOCATION** | western Africa |
| **STATUS** | Critically Endangered |

• December 17th •

## BANDED MONGOOSE

Unlike most mongooses, which are solitary, this one lives in a colony of up to 40 animals. Needing dens for shelter, they build tunnel systems with multiple entrances under termite mounds, bushes or in thickets on the savanna or in the grasslands. They sleep in a den overnight, change dens frequently and establish their territories by scent-marking. They search for food as individuals but will work as a team to kill if they encounter a venomous snake.

| | |
|---|---|
| **SCIENTIFIC NAME** | *Mungos mungo* |
| **ANIMAL GROUP** | Mammals |
| **LENGTH/WEIGHT** | up to 2.5 ft. including tail/6 lb. |
| **DIET** | carnivore: beetles, millipedes, other insects, frogs, snakes, eggs |
| **LOCATION** | central and southern Africa |
| **STATUS** | Least Concern |

• December 18th •

## LEOPARD SEAL

A fierce ambush hunter and the top predator off the shores of Antarctica, the leopard seal is only predated by the orca (*see p.197*). Its jaw can open 160 degrees to clamp its large, curving canine teeth down on a gentoo penguin with incredible force. Its rear teeth are used to filter krill out of the water. It finds its prey by sight and smell and can move in bursts of up to 25 mph.

| | |
|---|---|
| **SCIENTIFIC NAME** | *Hydrurga leptonyx* |
| **ANIMAL GROUP** | Mammals |
| **LENGTH/WEIGHT** | up to 11 ft./1,100 lb. |
| **DIET** | carnivore: penguins, krill, fish, birds, seal pups |
| **LOCATION** | Southern Ocean |
| **STATUS** | Least Concern |

• December 19th •

# SOUTHERN CASSOWARY

This is a flightless bird with nearly 5-inch-long claws and a very powerful, killing kick. If threatened, it can charge at up to 31 mph, sometimes jumping 5 feet in the air as part of the attack. The casque on its head is made of keratin, like human nails, and it acts as protection as the bird pushes through dense vegetation. The necks of its striped chicks begin to change color at about 6 to 9 months.

| | |
|---|---|
| **SCIENTIFIC NAME** | Casuarius casuarius |
| **ANIMAL GROUP** | Birds |
| **HEIGHT/WEIGHT** | up to 7 ft./170 lb. |
| **DIET** | omnivore: mainly fruit, also small rodents, snakes, insects, snails, fish |
| **LOCATION** | northern Australia, southeastern Asia |
| **STATUS** | Least Concern |

• December 20th •

# SOUTHERN SEA LION

These maned sea lions live along shorelines and beaches, the large bulls (males) patrolling their territories, in which there are usually around 18 females. They rest both during the day and at night, and can sleep equally well either in the water or on the beach.

| | |
|---|---|
| **SCIENTIFIC NAME** | Otaria flavescens |
| **ANIMAL GROUP** | Mammals |
| **LENGTH/WEIGHT** | up to 10 ft./770 lb. |
| **DIET** | carnivore: fish, squid, crustaceans, penguins |
| **LOCATION** | Atlantic and Pacific Oceans around southern South America |
| **STATUS** | Least Concern |

• December 21st •

# BANDED SEA KRAIT

This semiaquatic sea snake is highly venomous. It has a flattened tail which protects it because predators think it is the krait's head and will avoid it because they might get bitten. The adult snakes rest and nest on rocky headlands and beaches.

| | |
|---|---|
| **SCIENTIFIC NAME** | Laticauda colubrina |
| **ANIMAL GROUP** | Reptiles |
| **LENGTH/WEIGHT** | up to 5 ft./4 lb. |
| **DIET** | carnivore: eels, small fish |
| **LOCATION** | Indian and Pacific Oceans |
| **STATUS** | Least Concern |

• December 22nd •

# BEARDED BARBET

Named for its "beard" of black feathers, this bird has sawlike edges on its beak that help it cut through tough fruit. Like the woodpecker it is related to, it carves out holes in tree trunks to make its nests.

| | |
|---|---|
| **SCIENTIFIC NAME** | Pogonornis dubius |
| **ANIMAL GROUP** | Birds |
| **LENGTH/WEIGHT** | up to 10 in./3.7 lb. |
| **DIET** | omnivore: fruit, insects |
| **LOCATION** | western Africa |
| **STATUS** | Least Concern |

## • December 23rd •
## CARACAL

Leaping 10 feet into the air, this cat catches birds in flight with its hooked claws. It has very sensitive hearing to track prey on the ground. Its foot pads are cushioned by fur and make very little noise, so prey is not aware of its approach. If it needs to, the caracal can chase an animal down at up to 50 mph. Caracals live in many different habitats but are most at home somewhere that has both open savanna and trees for cover.

| | |
|---|---|
| **SCIENTIFIC NAME** | *Caracal caracal* |
| **ANIMAL GROUP** | Mammals |
| **LENGTH/WEIGHT** | up to 4 ft. including tail/40 lb. |
| **DIET** | carnivore: mongooses, rodents, antelope, monkeys |
| **LOCATION** | Africa, southwestern Asia |
| **STATUS** | Least Concern |

## • December 24th •
## BILBY

This burrowing marsupial is nocturnal. It has long ears like a rabbit and was once found across 70 percent of Australia, but now only lives in desert areas. It is about the size of a pet cat and digs large tunnel systems to protect itself from the heat and predators.

| | |
|---|---|
| **SCIENTIFIC NAME** | *Macrotis lagotis* |
| **ANIMAL GROUP** | Mammals |
| **LENGTH/WEIGHT** | up to 35 in. including tail/5 lb. |
| **DIET** | omnivore: bulbs, tubers, seeds, insects, fungi |
| **LOCATION** | Australia |
| **STATUS** | Vulnerable |

• December 25th •

Spending most of its time under crevices or logs where it can stay safe and moist, this salamander is brightly colored to warn off predators, and can spray defensive chemicals up to 6 feet if it is attacked. It is not active when there are extremes of heat and hibernates (*see pp. 174–175*) in winter. Traditional folk tales told of these salamanders being born from fire.

| | |
|---|---|
| **SCIENTIFIC NAME** | *Salamandra salamandra* |
| **ANIMAL GROUP** | Reptiles |
| **LENGTH/WEIGHT** | up to 1 ft./40 lb. |
| **DIET** | carnivore: worms, slugs, centipedes, flies, beetles |
| **LOCATION** | central and southern Europe |
| **STATUS** | Least Concern |

• December 26th •

# HONEY BADGER

Also known as the ratel, this is a member of the weasel family. It gets its name from raiding beehives, but it will eat many other things, including young crocodiles. It is fierce and able to take kills off larger animals, including black-backed jackals (*below*) and lions. It has very thick skin and sharp teeth, and likes to eat venomous snakes. It uses its long claws to dig the burrows where it rests.

| | |
|---|---|
| **SCIENTIFIC NAME** | *Mellivora capensis* |
| **ANIMAL GROUP** | Mammals |
| **LENGTH/WEIGHT** | up to 3.3 ft. including tail/35 lb. |
| **DIET** | omnivore: honey, insects, reptiles, small mammals, birds, plants |
| **LOCATION** | Africa, southern and southwestern Asia |
| **STATUS** | Least Concern |

• December 27th •

# NARWHAL

This "unicorn of the sea" is one of the deepest diving mammals, reaching depths of almost 6,000 feet to feed during the winter months. In the summer, they fish close to the shore. The males have a single spiral tusk, a straight tooth, up to 10 feet long, with which they compete for mates. Narwhals are classed as toothed whales because of the tusk but do not have teeth in their mouth, so they swallow fish whole.

| | |
|---|---|
| **SCIENTIFIC NAME** | *Monodon monoceros* |
| **ANIMAL GROUP** | Mammals |
| **LENGTH/WEIGHT** | up to 25 ft. including tusk/1.8 tons |
| **DIET** | carnivore: fish, shrimps, squid |
| **LOCATION** | Arctic Ocean |
| **STATUS** | Least Concern |

• December 28th •

# HARRIS HAWK

Uniquely in the bird world, these hawks stand on top of one another when they cannot find places to perch! They are social birds, hunting as a team and nesting together. They often surround or chase prey for the other hawks to catch. The group also shares nesting responsibilities. When a female is incubating eggs or caring for the chicks, other females help and defend the nest.

| | |
|---|---|
| **SCIENTIFIC NAME** | *Parabuteo unicinctus* |
| **ANIMAL GROUP** | Birds |
| **WINGSPAN/WEIGHT** | up to 3.9 ft./2.2 lb. |
| **DIET** | carnivore: hares, rabbits, quails, reptiles |
| **LOCATION** | North, Central and South America |
| **STATUS** | Least Concern |

• December 29th •

# PEACOCK

Also known as the Indian peafowl, this wonderful bird has the most superb method of attracting a mate. Peahens choose the male that has the most eyespots on their feathers, so the males take three years to grow their tails. They display by spreading the tail feathers high and wide and "wing-shaking" to make the colors shine when the sun catches them at different angles.

| | |
|---|---|
| **SCIENTIFIC NAME** | *Pavo cristatus* |
| **ANIMAL GROUP** | Birds |
| **LENGTH/WEIGHT** | up to 5 ft./13 lb. |
| **DIET** | omnivore: insects, worms, lizards, termites, flowers, grain, grass, bamboo shoots |
| **LOCATION** | Asia |
| **STATUS** | Least Concern |

• December 30th •

# RAINBOW LIZARD

When startled, the colors of this lizard become brighter. It has long, powerful hind legs to leap or run away quickly from a predator. Also known as the common agama, this lizard lives in dry forests, grasslands and deserts, where it ambushes prey. It has large front teeth and strong jaws for larger animals, and a sticky tongue to sweep up insects.

| | |
|---|---|
| **SCIENTIFIC NAME** | *Agama agama* |
| **ANIMAL GROUP** | Reptiles |
| **LENGTH/WEIGHT** | up to 1 ft. including tail/2.2 lb. |
| **DIET** | omnivore: insects, small mammals, reptiles, fruit, grasses, flowers |
| **LOCATION** | eastern Africa |
| **STATUS** | Least Concern |

## • December 31ˢᵗ •
## ASIAN ELEPHANT

Communicating with low-pitched sounds that humans can hardly hear, a herd of these forest elephants usually numbers six to seven adults and their young. The adults are related females led by a matriarch, the oldest female. The bulls are usually solitary. With smaller bodies and ears than the African elephant (*see p.118*), these elephants pass their time finding and eating up to 285 pounds of vegetation each in a single day.

| | |
|---|---|
| **SCIENTIFIC NAME** | *Elephas maximus* |
| **ANIMAL GROUP** | Mammals |
| **LENGTH/WEIGHT** | up to 26 ft. including tail/7 tons |
| **DIET** | herbivore: tree bark, roots, leaves, grasses, bamboo |
| **LOCATION** | southern Asia |
| **STATUS** | Endangered |

# IN DANGER

An animal is described as endangered when there are so few of its species left that it is in danger of becoming extinct. Scientists need to understand what threats animals face around the world and figure out how to prevent this happening.

## PHILIPPINE EAGLE (SEE P.65)

**Threats**: *deforestation, hunting, traps*
It is estimated that there are now fewer than 400 nesting pairs of these eagles, making it one of the rarest birds in the world. Loss of habitat has resulted in the loss of nesting sites. Illegal hunting by farmers to protect their livestock and accidental capture in traps set for wild pigs have also done enormous damage.

## ZEBRA SHARK (SEE P.18)

**Threats**: *fishing, threats to habitat*
This is one of the sharks that is killed for its fins for soup, despite many countries banning the trade. It is also fished for meat and its liver eaten for its vitamins. Its habitat is threatened by pollution and it is virtually extinct in areas it used to live.

## BLUE POISON FROG (SEE P.34)

**Threats**: *deforestation, disease, pet trade*
Deforestation carried out to build houses and establish farms has devastated the rainforests that were home to this little frog. Unfortunately, many of the species have died, and are dying, from a deadly fungal disease as well. To add to this, they are being taken from their habitat to be sold as pets.

## KING COBRA (SEE P.48)

**Threats**: *deforestation, harvesting, killing*
The forests the cobra lives in are being destroyed for logging and agriculture. The skin and other parts of its body are harvested for fashion and medicine. And many are simply killed because people are afraid of them.

## POLAR BEAR (SEE P.15)

**Threats**: *climate change, pollutants, oil exploration*
The melting of sea ice in the Arctic means that the bear does not have access to seals, its main prey. It is also prey to pollutants and disease in the food it does find. Gas and oil exploration exposes polar bears to contaminating spills or leaks as well.

## JEWEL BEETLE (SEE P.35)

**Threats**: *commercial and illegal logging*
This beautiful beetle needs dead or dying trees because that is where it lays its eggs. With so many forests across Europe being cut down to make money, fewer trees are available, and the beetle is becoming extinct in many countries.

## BLACK RHINO (SEE P.130)

**Threats**: *illegal trade in rhino horn, habitat loss*
Despite all efforts, poachers continue to illegally kill rhinos for their horns. The horns are used for traditional medicine, mainly in Asia, and are considered a symbol of status and wealth. These animals also suffer because the grasslands they live on have been converted into land for farming or to build houses.

## GREEN SEA TURTLE (SEE P.140)

**Threats**: *harvesting of eggs, loss of beaches, bycatch*
Turtles return to the same beach every year to lay their eggs in the sand. People take the eggs for food, while tourism and coastal development have caused loss of this nesting habitat. At sea, the turtles become bycatch – they are caught and drowned in both commercial and recreational fishing nets. They are also poached illegally, to eat, or for their skin and shells.

## GIANT SALAMANDER (SEE P.29)

**Threats**: *habitat destruction, water pollution, killed for food*
The building of dams and changes in the way rivers flow have destroyed much of the salamander's habitat and introduced it to pollution and disease. The largest danger, however, is exploitation for food, with wild populations being taken so they can be farmed for sale.

## ARMORED SNAIL (SEE P.68)

**Threats**: *altered water chemistry, limited habitat*
This extraordinary snail is considered most in danger from deep sea mining and exploration. Every intrusion into their very limited habitat near hydrothermal vents introduces chemical changes to the water.

# CONSERVATION SUCCESS

In the battle to protect animals, there are some real success stories, either through the work of individuals, or the collaboration of wildlife organizations and governments around the world. Here are just a few of those stories.

## RED-CROWNED CRANE (SEE P.127)

On the island of Hokkaido in Japan, the population of red-crowned cranes was taken to the brink of extinction by hunting and by farmland replacing their habitat. Among the measures that have saved them, local farmers now feed the cranes corn and buckwheat to help them increase their numbers.

## SUMATRAN ORANGUTAN (SEE P.134)

On the island of Sumatra, this great ape was threatened by palm oil plantations wiping out huge areas of rainforest. The young were also taken to be sold as pets. The orangutan is still considered Critically Endangered because because two thirds of the population do not live in protected areas of the island. However, today there are dedicated groups and individuals helping to repopulate the island with orangutans.

## JAGUAR (SEE P.116)

Wild jaguars were locally extinct in the Ibera wetlands of Argentina 70 years ago. In 2021, rewilding efforts led to the release of a female and her two cubs. This was followed by the release of more females and cubs and, finally, a male in 2022. This is the start of a repopulation of the area.

## BALD EAGLE (SEE P.103)

Only 487 nesting pairs remained in the world by 1963, but in 1972 the harmful pesticide DDT, which had wiped out the bird in 48 states, was banned in the United States. In 1973, the Endangered Species Act, which mentions the bald eagle in particular, was brought in. By 1995, it had moved from IUCN Endangered status to Threatened, and today there are 71,400 nesting pairs.

## BLACK RHINO (SEE P.130)

In Kenya, rhinos are illegally slaughtered for their horns. In 2015 alone, 1,400 rhinos were killed across Africa, and many rangers employed to protect the wildlife lost their lives. Today, rangers and communities work together, and helicopters are used to reach remote places quickly. The rangers are also able to monitor the animals' movements with trackers and cameras, and in 2020 not a single rhino was poached.

## PANDA (SEE P.95)

In the 1970s, there were only 1,000 pandas left in the wild, and research and satellite imagery in the 1980s showed that their habitat had reduced by at least 50 percent. In the 1990s, plans were put in place to protect these spectacular animals, and by 2014 the number of pandas had increased to 1,864.

## HUMPBACK WHALE (SEE P.74)

By 1950, it was thought that there were only around 5,000 humpbacks left from a population that once numbered 125,000. In 1966, an international law banning commercial whaling was brought in and was the beginning of change. Today, the global population of mature humpbacks is estimated at more than 84,000, with numbers increasing.

# QUIZ

Now that you've met some amazing animals, can you answer these quiz questions about them? On the opposite page, you can also find some puzzles to solve. To check your answers, or if you get stuck, look at the upside-down text at the bottom.

**1. IN WHICH COUNTRY COULD YOU FIND THE KOALA, RED KANGAROO AND RAINBOW LORIKEET?**

a) USA
b) Australia
c) South Africa

**2. SACRED SCARAB BEETLES CAN ROLL BALLS OF DUNG UP TO 50 TIMES HEAVIER THAN THEIR BODY WEIGHT.**

True or false?

**3. WHICH OF THESE ANIMALS DOES NOT ROLL ITSELF INTO A BALL AS PROTECTION FROM PREDATORS?**

a) Pangolin
b) Four-toed hedgehog
c) Thorny devil

**4. THE NORTHERN ROCKHOPPER PENGUIN IS BIGGER THAN THE EMPEROR PENGUIN.**

True or false?

**5. WHICH OF THESE ANIMALS CHANGES COLOR TO CAMOUFLAGE ITSELF, FOR DISGUISE OR TO DEFEND ITS TERRITORY?**

a) Groundhog
b) Panther chameleon
c) Red-crowned crane

**6. THE SIDEWINDER, RETICULATED PYTHON AND KING COBRA ARE ALL TYPES OF WHICH ANIMAL?**

a) Fish
b) Butterfly
c) Snake

**7. WHICH OF THESE ANIMALS COULD YOU NOT FIND IN THE ARCTIC?**

a) Polar bear
b) Narwhal
c) Fennec fox

**8. WHAT DO LEATHERBACK TURTLES, SOCKEYE SALMON, CANADA GEESE AND RUBY-THROATED HUMMINGBIRDS ALL HAVE IN COMMON?**

a) They all go on long-distance migrations
b) They are all carnivores
c) They are all endangered

# PUZZLES

1. EACH OF THESE FIVE RAINFOREST-DWELLING ANIMALS HAS BEEN DISGUISED, WITH ITS FIRST AND LAST LETTERS REMOVED. CAN YOU WORK OUT WHAT THE MISSING LETTERS ARE TO REVEAL THE ANIMALS? ALL OF THE MISSING LETTERS ARE LISTED BELOW TO HELP YOU.

_ ANDRIL _
_ OAL _
_ ONOB _
_ EOPAR _
_ RANGUTA _

**A L O M N K O B D L**

BONUS QUESTION: WHICH THREE OF THESE RAINFOREST ANIMALS ARE TYPES OF MONKEY OR APE?

2. CAN YOU WORK OUT WHICH COLORFUL STRIPED FISH IS HIDING IN THIS ANAGRAM? UNSCRAMBLE THE LETTERS TO REVEAL THE ANIMAL.

**H L F O C I S N W**

---

## PUZZLES

1. MANDRILL
   KOALA
   BONOBO
   LEOPARD
   ORANGUTAN

   Bonus question: MANDRILL, BONOBO, ORANGUTAN

2. CLOWNFISH

## QUIZ

1. B - Australia
2. True! You can find out more about these amazing animals on page 82.
3. C - Thorny devil. Head to page 91 to learn more.
4. False. The emperor penguin, found on pages 38-39, is the largest type of penguin.
5. B - Panther chameleon. You can find it on page 114, this time not camouflaged!
6. C - They are all types of snake
7. C - Fennec fox
8. A - they all go on long-distance migrations. You can find them all on pages 84-85.

215

# GLOSSARY

**APEX PREDATOR**  An animal at the very top of the food chain in a particular environment. The animal hunts other animals but is not hunted itself.

**BALEEN**  Fringed plates that hang down like a curtain from some whales' upper jaws and are used to filter small food items from water.

**BIOLUMINESCENCE**  The production of light by animals to light up dark places, found in fish and insects as well as simple animals that live in the sea, such as blobfish.

**BIVALVE**  A type of mollusk that has a hinged shell composed of two halves.

**BRACHIATE**  To swing by the arms from branch to branch when traveling through trees, the way apes such as gibbons and orangutans do.

**BREACH**  To leap out of the water into the air and splash back down again. Animals that breach include humpback whales and great white sharks.

**BREEDING**  Producing offspring by mating.

**BRUMATION**  Dormancy for cold-blooded animals. Frogs, lizards, snakes and turtles brumate.

**CAMOUFLAGE**  The way the color or shape of some animals allows them to blend in with their surroundings, either to ambush or avoid other animals.

**CARRION**  The remains of a dead animal.

**CEPHALOPODS**  A group of marine mollusks that includes squid, cuttlefish and octopuses.

**CETACEANS**  Marine mammals that include whales, dolphins and porpoises.

**COCOON**  A covering that protects an animal. A spider spins a silk cocoon to protect itself or its eggs.

**COLD-BLOODED**  Describes an animal that cannot produce its own body heat and relies on the environment to regulate its body temperature. Reptiles, fish, amphibians and invertebrates are cold-blooded.

**COLONY**  A large group of the same species of animal living together, for example ants or termites.

**CONSERVATION**  The active protection of animals and their habitat.

**CONSTRICTOR**  Describes a snake that coils around its prey and crushes it.

**CRUSTACEAN**  An invertebrate with a exoskeleton and jointed limbs, usually found in water.

**ECHOLOCATION**  A way some animals, such as bats, sense objects and prey. They produce bursts of sound that bounce off objects and echo back.

**ESTIVATION**  Similar to hibernation, this is when animals slow their activity to stay cool in hot or dry weather.

**EXOSKELETON**  The external skeleton, or hard outer shell, that supports and protects the bodies of many invertebrates, such as beetles and lobsters.

**EXTINCT**  No longer existing or living.

**FILTER-FEED**  To eat by sieving food from water.

**GILLS**  Frilly or feathery body parts on the outside of water-living animals that are used to collect oxygen to breathe underwater and get rid of carbon dioxide.

**HERMAPHRODITE**  An animal that has both male and female reproductive organs.

**HIBERNATION**  When a warm-blooded animal slows its heart rate to save energy and survive cold weather without eating much. Some go into a deep sleep, while others slow down in a sleeplike state called torpor.

**INCUBATION**  Sitting on eggs to hatch them.

**IRIDESCENT**  Having luminous colors that appear to change when seen from different angles.

**KERATIN**  The substance that makes up the skin, hair, nails, horns, hooves, beaks and feathers of animals.

**KRILL**   Shrimplike crustaceans that live in the ocean.

**LARVA**   (plural: larvae) Describes an insect after it hatches from an egg and before it changes into its adult form. Larvae do not have wings and look like worms. A caterpillar is the larval stage of a butterfly.

**MARSUPIAL**   Mammals, including kangaroos and opossums, that have pouches to carry young.

**METAMORPHOSE**   To change in body shape, for example when a caterpillar metamorphoses into a butterfly.

**MIGRATION**   A long journey to a different area to search for food, find a place to breed or escape from cold weather.

**MIMICRY**   The copying of another animal, usually in order to deter or scare off predators.

**MOLLUSKS**   Invertebrates that have soft bodies. Many of them have protective shells.

**MONOTREMES**   A group of egg-laying mammals that consists of echidnas and platypuses.

**MUCUS**   A slimy liquid produced by some animals, for example snails, that use it to slide along the ground.

**NEMATOCYST**   One of the tiny, stinging organs of a jellyfish, coral or sea anemone.

**NOCTURNAL**   Describes an animal that is active mainly at night and sleeps during the day.

**NYMPH**   The young stage of some kinds of invertebrates, especially insects such as dragonflies. A nymph looks like a tiny version of the adult but does not have wings yet.

**PARASITE**   An animal that lives in or on another living animal, either feeding on them or on the food they eat.

**PLANKTON**   Microscopic living animals and plants that drift near the surface of the water.

**POLLINATE**   The act of moving pollen grains from one flowering plant to another, for example by bees. This action fertilizes plants and helps them make seeds.

**PREDATOR**   An animal that hunts and eats other animals.

**PREHENSILE**   Describes something that can wrap around an object and grip it. Some monkeys have prehensile tails.

**PREY**   An animal hunted and eaten by another animal.

**PRIMATES**   A group of mammals that includes humans, apes and monkeys.

**RODENTS**   A group of small, gnawing mammals that have a single pair of incisors. Rodents include rats, squirrels and beavers.

**SCAVENGER**   An animal that searches for and eats dead animals left by predators.

**SCENT-MARKING**   The spraying by an animal of its territory with a strong-smelling substance to keep out intruders.

**SPECIES**   A group of living things that look alike and can breed together to produce young that resemble their parents.

**STRIDULATION**   The production of sound by rubbing together two parts of the body. Grasshoppers and crickets stridulate.

**SYMBIOTIC**   Describes a partnership between animals of two different species, in which both gain from the relationship.

**TALON**   A large, hooked claw of a bird of prey.

**TENTACLES**   The long, fleshy feelers that some animals, such as octopuses, use to catch prey.

**TOXIN**   Poison that can harm living things.

**VENOMOUS**   Describes an animal that injects toxins in defense or to kill prey.

**VERTEBRAE**   (singular: vertebra) The bones that link together to form the backbone of vertebrates.

**VOCAL SAC**   The throat pouch of male frogs and toads that inflates to magnify the sounds that they make.

**WARM-BLOODED**   Describes the ability of an animal to regulate body temperature independently of the surroundings. Mammals and birds are warm-blooded.

# INDEX

## A
aardvark 96
aardwolf 138
achallo 156
African fish eagle 27
African hermit spider 138
African savanna elephant 118
African wild dog 152
alligator snapping turtle 147
alpine ibex 197
alpine newt 12, 75
Altiplano chinchilla mouse 156
Amazon leaf fish 131
Amazon river dolphin 99
Amazon tree boa 139
American alligator 56
American bison 12, 199
amphibian, largest 29
amphibians 12
Andean cock-of-the-rock 16
Andean condor 123
Andean night monkey 164
animal assassins 146–147
Anna's eighty-eight butterfly 92
ant bear 188
apex predators 79
Arabian oryx 91
Arctic fox 21
Arctic hare 154
Arctic tern 28
armadillo girdled lizard 24
armored ground cricket 18
armored snail 68, 211
Asian elephant 209
Asian garden dormouse 64

Asiatic black bear 175
Atlantic puffin 182
Australian hornet 104
Australian walking stick 191
axolotl 59
aye-aye 66

## B
bald eagle 12, 103, 212
banded archerfish 20
banded demoiselle 101
banded mongoose 202
banded sea krait 204
barn owl 67
barn swallow 42
baya weaver bird 104
bearded barbet 204
bearded vulture 193
beluga sturgeon 13, 99
beluga whale 155
Bennett's flying fish 12, 123
Bhutan glory 194
big dipper firefly 164
bigfin reef squid 17
bilby 205
birds 12
black-bearded jackal 206
black-bellied hamster 121
black-crowned squirrel monkey 128
black-legged kittiwake 45
black mamba 44
black panther 147
black rhino 130, 211
black widow spider 173
blackfin icefish 155
blobfish 68
blue emperor butterfly 129
blue-footed booby 107
blue jay 96
blue mountain swallowtail 129
blue poison frog 34, 210
blue sea dragon 58
blue-tailed bee-eater 81
blue whale 106

blue wildebeest 23
bongo 51
bonobo 51
brown fur seal 80
brown pelican 137
brown recluse spider 42
brown-throated sloth 57
brumation 175
Budgett's frog 152
Burmese python 116
burrowing 105, 186–187
burrowing owl 186

## C
Cambodian blue-crested agama 12
Canada goose 84
cane toad 120
Cape fur seal 80
capybara 97
caracal 205
Caribbean reef octopus 13, 141
caribou 155
central coast stubfoot toad 145
cheetah 29
chimpanzee 88
Chinese giant salamander 29, 211
Chinese water deer 25
classifications 13
claws 56–57
clownfish 109
coconut crab 106
cold-blooded animals 12
common agama 208
common cuttlefish 198
common house gecko 43
common kingfisher 99
common marmoset 198
common opossum 196
common pipistrelle 43
common pondskater 98
common shrew 60
common stingray 54
common wombat 186
conservation 13, 212–213

copperband butterflyfish 139
coral reefs 140–141
cotton harlequin beetle 66
crested porcupine 40
crested serpent eagle 51
crucifix frog 58
cuttlefish, common 198

### D

daisy parrotfish 97
deforestation 210–211
desert kangaroo rat 71
desert pocket gopher 187
devil scorpion 90
diamondback terrapin 170
disease 210
dormancy 174–175
duck-billed platypus 125

### E

eastern box turtle 174
eastern coral snake 25
eastern hognose 77
eastern meadowlark 124
eastern water-holding frog 174
electric eel 20
elephant shrew 109
elf owl 41
elk 26
emerald tree boa 75
emerald tree monitor 133
emperor penguin 38
emperor scorpion 13, 157
emperor tamarin 161
endangered animals 210–211
Española giant tortoise 111
estivation 174
Eurasian badger 144
Eurasian green woodpecker 18
Eurasian otter 98
European fire-bellied toad 12, 161
European green toad 136
European hamster 121
European mole 113
European rabbit 113

extinction 210, 212
exoskeleton 13

### F

false coral snake 145
false tomato frog 61
fat-tailed dwarf lemur 175
fennec fox 91
feral pigeon 181
fiddler crab 56
fire salamander 206
fish 12
fishing cat 17
flat bark beetle 132
flight 122–123
flying dragon lizard 122
flying lemur 171
four-toed hedgehog 148
fried egg jellyfish 191
frill-necked lizard 114
frilled shark 69

### G

Galápagos pink land iguana 37
giant anteater 188
giant armadillo 57
giant clam 129
giant day gecko 128
giant forest ant 83
giant oceanic manta ray 168
giant panda 95, 212
giant squid 34
giant tiger land snail 160
gila monster 108
giraffe weevil 120
gladiator spider 20
glass-winged butterfly 65
goblin shark 116
golden-spotted tiger beetle 170
goliath frog 16
goliath tarantula 188
Gouldian finch 156
gray heron 98
gray long-eared bat 123
gray whale 85
gray wolf 172
great blue-spotted mudskipper 171
great bustard 130

great hammerhead shark 54
great hornbill 178
great spotted woodpecker 59
great white shark 79
greater blue-ringed octopus 146
greater roadrunner 190
green anaconda 179
green sea turtle 12, 140, 211
grizzly bear 87
ground pangolin 89
groundhog 154
gulper eel 117

### H

habitat destruction 210, 211
Hainan gibbon 133
hairy frogfish 64
harbour porpoise 199
harpy eagle 56
Harris hawk 207
Hawaiian hoary bat 36
Hermann's tortoise 33
hibernation 174–175
hippopotamus 117
honey badger 206
honey bear 133
hooded pitohui 146
hoopoe 24
house mouse 43
humpback anglerfish 69
humpback whale 74, 213
hunting methods 20–21

### I

ili pika 169
immortal jellyfish 197
Indian bullfrog 183
Indian flying fox 27
Indian peafowl 208
Indo-Pacific bottlenose dolphin 141
Indo-Pacific sailfish 28
Indus Valley frog 183
invertebrates 12, 13

## J

jaguar 116, 212
jaws 116, 117
jewel beetle 35, 211
joro spider 67

## K

Kaiser's mountain newt 89
king cobra 48, 210
kingfisher, common 99
kinkajou 133
koala 29
komodo dragon 189

## L

lammergeier 193
lar gibbon 82
large-eared garden dormouse 64
largetooth sawfish 190
laughing kookaburra 129
Laysan albatross 28
leaf-curling spider 105
leaf-cutter ant 183
leaf-nosed moray eel 52
leafy seadragon 33
least chipmunk 32
least weasel 187
leatherback turtle 85
leopard 57
leopard seal 202
leopard shark 149
lesser flamingo 37
lifespan, longest 28
lion 22
long-nosed horned frog 108
longhorn cowfish 25
longspine porcupine fish 179
luna moth 72

## M

Macleay's spectre 191
mammals 12
mandarin duck 180
mandrill 101
maned wolf 47
margay 21
marine iguana 177
marmoset, common 198

mason wasp 104
meerkat 93
Mexican prairie dog 105
migration 28, 84–85
mimic octopus 77
mimicry 76–77
monarch butterfly 13, 84
monitor lizard 189
montane hourglass tree frog 83
moose 26
mountain gazelle 136
mute swan 100

## N

Namid darkling beetle 90
narwhal 207
neon tetra 121
nests 104–105
new species 13
nightjar 165
nightlife 164–165
Nile crocodile 80
nocturnal animals 164–165
North American beaver 104
North Island brown kiwi 173
northern cardinal 115
northern giraffe 119
northern raccoon 42
northern red salamander 61
northern rockhopper penguin 93
nudibranch 58

## O

oceans, deep 68–69
ocelot 40
ogre-faced spider 20
okapi 163
opossum, common 196
orca 197
orchid mantis 21
ornate bluet damselfly 60
ostrich 12, 166
owl butterfly 76

## P

painted anemone 152
painted dog 152
palmate newt 125
panther chameleon 114
paradise tree snake 122
peacock 208
peacock mantis shrimp 72
peacock worm 92
peregrine falcon 181
Philippine eagle 65, 210
pine marten 73
pipistrelle, common 43
pistol shrimp 196
plains zebra 195
poaching 210, 211
polar bear 15, 210
pollutants 210
polyphemus moth 165
Pompeii worm 36
Portugese man o' war 19
potter wasp 104
Potter's angelfish 140
pronghorn 44
pygmy hippopotamus 45

## Q

Queen Alexandra's birdwing 49

## R

Raggiana bird-of-paradise 63
rainbow lizard 208
rainbow lorikeet 88
ratel 206
raven 154
red-bellied piranha 50
red-crowned crane 127, 212
red deer 184
red-eyed tree frog 49
red fox 181
red kangaroo 159
red lionfish 137

red panda  195
red-sided garter snake  175
red squirrel  132
red-tailed bumble bee  174
reef stonefish  92
reindeer  155
reptiles  12
resplendent quetzal  145
reticulated python  50
ribbon eel  52
ring-tailed lemur  53
robber fly  146
rock dove  181
Rondo bushbaby  165
Rondo dwarf galago  165
ruby-throated hummingbird  84

## S

sacred scarab beetle  82
saltwater crocodile  31
scarlet macaw  112
screaming eagle  27
sea eagle  103
sea urchin  186
secretary bird  153
seven-spot ladybird  124
Seychelles leaf insect  76
shoebill  169
short-horned grasshopper  13, 108
shortfin mako  196
shrew, common  60
sidewinder  70
sika deer  180
snail kite  32
snowy owl  150
sockeye salmon  85
southern cassowary  203
southern flying squirrel  122
southern sea lion  204
spectacled caiman  12, 53
speed  28, 29
sphinx hawk moth  77
spinner dolphin  143

spiny lobster  162
spotted eagle ray  157
spotted hyena  117
Steller's jay  73
striped skunk  182
Sumatran orangutan  134, 213
Sumatran tiger  135
sun bear  115
sunda colugo  171
superb lyrebird  76
superpowers  196–197
swallowtail butterfly  194

## T

talons  56–57
tardigrade  36
Tasmanian devil  164
tawny frogmouth  52
teeth  116–117
Texas horned lizard  61
thorny devil  91
tiger quoll  50
timber rattlesnake  147
toco toucan  194
tokay gecko  101
Tonkin bug-eyed frog  124
torpor (semi-hibernation)  32
Townsend's big-eared bat  12, 148
tree dwellers  132–133
treecreeper  132
tropical rainforests  50–51
tufted gray langur  180
tunki  16

## U

unicorn of the sea  207
urban environments  180–181

## V

vampire ground finch  37
vampire squid  69
vertebrates  12
vicuña  160
Vietnamese giant centipede  41
Vietnamese mossy frog  124
violin spider  42
Vogelkop bowerbird  105
volcanoes  36–37

## W

warm-blooded animals  12
waterbear  36
webs  20, 67, 105, 138, 173, 187
western honeybee  191
western lowland gorilla  200
whale shark  141
white-tufted marmoset  198
white-winged petrel  149
wild bactrian camel  90
wild boar  178
wombat, common  186

## Y

yellow assassin fly  146
yellow-banded sweetlips  12, 83
yellow-naped parrot  162
yeti crab  68

## Z

zebra seahorse  140
zebra shark  18, 210